SpringerBriefs in Physics

W0225821

Series Editors

Balasubramanian Ananthanarayan, Centre for High Energy Physics, Indian Institute of Science, Bangalore, Karnataka, India

Egor Babaev, Department of Physics, Royal Institute of Technology, Stockholm, Sweden

Malcolm Bremer, H. H. Wills Physics Laboratory, University of Bristol, Bristol, UK

Xavier Calmet, Department of Physics and Astronomy, University of Sussex, Brighton, UK

Francesca Di Lodovico, Department of Physics, Queen Mary University of London, London, UK

Pablo D. Esquinazi, Institute for Experimental Physics II, University of Leipzig, Leipzig, Germany

Maarten Hoogerland, University of Auckland, Auckland, New Zealand

Eric Le Ru, School of Chemical and Physical Sciences, Victoria University of Wellington, Kelburn, Wellington, New Zealand

Dario Narducci, University of Milano-Bicocca, Milan, Italy

James Overduin, Towson University, Towson, MD, USA

Vesselin Petkov, Montreal, QC, Canada

Stefan Theisen, Max-Planck-Institut für Gravitationsphysik, Golm, Germany

Charles H. T. Wang, Department of Physics, University of Aberdeen, Aberdeen, UK

James D. Wells, Department of Physics, University of Michigan, Ann Arbor, MI, USA

Andrew Whitaker, Department of Physics and Astronomy, Queen's University Belfast, Belfast, UK

SpringerBriefs in Physics are a series of slim high-quality publications encompassing the entire spectrum of physics. Manuscripts for SpringerBriefs in Physics will be evaluated by Springer and by members of the Editorial Board. Proposals and other communication should be sent to your Publishing Editors at Springer.

Featuring compact volumes of 50 to 125 pages (approximately 20,000–45,000 words), Briefs are shorter than a conventional book but longer than a journal article. Thus, Briefs serve as timely, concise tools for students, researchers, and professionals.

Typical texts for publication might include:

- A snapshot review of the current state of a hot or emerging field
- A concise introduction to core concepts that students must understand in order to make independent contributions
- An extended research report giving more details and discussion than is possible in a conventional journal article
- A manual describing underlying principles and best practices for an experimental technique
- An essay exploring new ideas within physics, related philosophical issues, or broader topics such as science and society

Briefs allow authors to present their ideas and readers to absorb them with minimal time investment. Briefs will be published as part of Springer's eBook collection, with millions of users worldwide. In addition, they will be available, just like other books, for individual print and electronic purchase. Briefs are characterized by fast, global electronic dissemination, straightforward publishing agreements, easy-to-use manuscript preparation and formatting guidelines, and expedited production schedules. We aim for publication 8–12 weeks after acceptance.

V. Parameswaran Nair

Geometric Quantization and Applications to Fields and Fluids

 Springer

V. Parameswaran Nair
Department of Physics
City College of New York
New York, NY, USA

ISSN 2191-5423   ISSN 2191-5431 (electronic)
SpringerBriefs in Physics
ISBN 978-3-031-65800-6   ISBN 978-3-031-65801-3 (eBook)
https://doi.org/10.1007/978-3-031-65801-3

This ebook was published Open Access with funding support from the Sponsoring Consortium for Open Access Publishing in Particle Physics (SCOAP3).

© The Editor(s) (if applicable) and The Author(s) 2024. This book is an open access publication.

Open Access This book is licensed under the terms of the Creative Commons Attribution 4.0 International License (http://creativecommons.org/licenses/by/4.0/), which permits use, sharing, adaptation, distribution and reproduction in any medium or format, as long as you give appropriate credit to the original author(s) and the source, provide a link to the Creative Commons license and indicate if changes were made.
The images or other third party material in this book are included in the book's Creative Commons license, unless indicated otherwise in a credit line to the material. If material is not included in the book's Creative Commons license and your intended use is not permitted by statutory regulation or exceeds the permitted use, you will need to obtain permission directly from the copyright holder.
The use of general descriptive names, registered names, trademarks, service marks, etc. in this publication does not imply, even in the absence of a specific statement, that such names are exempt from the relevant protective laws and regulations and therefore free for general use.
The publisher, the authors and the editors are safe to assume that the advice and information in this book are believed to be true and accurate at the date of publication. Neither the publisher nor the authors or the editors give a warranty, expressed or implied, with respect to the material contained herein or for any errors or omissions that may have been made. The publisher remains neutral with regard to jurisdictional claims in published maps and institutional affiliations.

This Springer imprint is published by the registered company Springer Nature Switzerland AG
The registered company address is: Gewerbestrasse 11, 6330 Cham, Switzerland

If disposing of this product, please recycle the paper.

To my sons Nityan and Haris

Preface

The fundamental dynamical variables of any physical system take values in what is referred to as the phase space. The geometry and topology of this space play a guiding role in the dynamics of the physical system. While this was well-appreciated and well-understood in classical dynamics, early formulations of quantum mechanics did not have an easy flexibility to accommodate features of geometry and topology. Over the years, this problem was addressed with increasing levels of sophistication. Geometric quantization gives an elegant framework for accommodating geometrical and topological features of the phase space. By now there are many books and mathematically sophisticated reviews of this topic. Most of these focus on the formalism and some of the subtleties involved. While this is of great value, I think that highlighting a variety of diverse applications, especially those which are physically motivated and interesting, can be a very useful complementary approach. This book is an attempt in this direction. In keeping with this motivation, most of the material here is presented from a physicist's point of view.

Some of the topics were covered in lectures at different summer schools in theoretical physics. More recently, a skeletal draft of most of the topics was prepared for lectures at the Second Autumn School on High Energy Physics & Quantum Field Theory in Yerevan, Armenia, in October 2014. This book is an augmented and updated version of the lecture notes.

I thank all the organizers of the summer school in Armenia for the invitation to speak there. I have discussed some of these topics with my colleagues and express my thanks for their insights and comments. I also thank my wife Dimitra for collaborations, discussions and for a span of time free from mundane worries while working on this. This work was supported in part by the U.S. National Science Foundation Grant PHY-2112729.

New York, USA   V. Parameswaran Nair
May 2024

Contents

Chapter 1
Introduction

A physical theory, as a logical explanation of physical phenomena, is to be constructed taking account of general principles and incorporating data and information from experiments. Any effects we attribute to the quantum nature of phenomena should be included from the outset. A classical description may then be obtained, in a suitable regime of parameters, as a useful and simpler working approximation. The flow of logic should thus be:

$$\left.\begin{array}{c}\text{General principles} +\\ \text{experimental input}\end{array}\right\} \implies \text{Quantum theory} \implies \text{Classical approximation.}$$

But the build-up of a theory along these lines is almost never done in practice. Primarily, this is because, at the human level of direct experience, most phenomena are well described by classical dynamics, and hence our intuition about physical systems is mostly classical. So we tend to start there and try to "quantize" the classical theory. This is a process with many ambiguities, but over the course of many years, we have learned to understand the structure of this procedure of quantization. In this book, we will attempt to describe some aspects of geometric quantization and consider a few examples or applications.

We will begin with some general observations on why we need such a procedure as geometric quantization. This is best illustrated by an example. Consider the elementary quantum mechanics of a single particle in three spatial dimensions. The operators of position $\hat{x}_i$, $i = 1, 2, 3$ and momentum $\hat{p}_i$ obey the Heisenberg algebra

$$\begin{aligned}\hat{x}^i \hat{x}^j - \hat{x}^j \hat{x}^i &= 0\\ \hat{x}^i \hat{p}_j - \hat{p}_j \hat{x}^i &= \mathrm{i}\delta^i{}_j\\ \hat{p}_i \hat{p}_j - \hat{p}_j \hat{p}_i &= 0\end{aligned} \tag{1.1}$$

As is well known these have the standard Schrödinger representation on the x-diagonal wave functions $\psi(x)$,

© The Author(s) 2024
V. P. Nair, *Geometric Quantization and Applications to Fields and Fluids*,
SpringerBriefs in Physics, https://doi.org/10.1007/978-3-031-65801-3_1

$$\hat{x}^i \, \psi = x^i \, \psi, \qquad \hat{p}_i \, \psi = -i \frac{\partial}{\partial x^i} \, \psi \qquad (1.2)$$

Notice that the commutation rules (1.1) and the specific representation (1.2) are expressed in terms of Cartesian coordinates. While we know that we should have the freedom of choosing any set of coordinates for the classical description, constructing the commutation rules and the operators in coordinate systems other than the Cartesian one is not straightforward. What is usually done in textbook solutions, say, of the Hydrogen atom in spherical polar coordinates is to set up the quantum theory and the Schrödinger equation first in the Cartesian basis (with $\hat{p}^2 = -\nabla^2$) and then make a change of coordinates. While this is an adequate working procedure for many situations, it is clearly unsatisfactory; one would like a procedure that works directly without the crutch of the Cartesian system. Also, in situations where we may have a curved space or a curved phase space, a quantization procedure which takes account of the geometry of the manifold is not just a desirable choice, but is actually needed.

There are also situations, such as in field theory, where the dynamical variables are components of fields and have no obvious Cartesian-like structure. In assigning commutation rules to the components of fields, a more general procedure is then called for. Geometric quantization is a partial answer to these concerns. It highlights the geometry and topology of the phase space and gives insights into many physical situations. But as it stands, it is still not a complete answer to the issues mentioned above. We will comment on some of these inadequacies later in the text.

There are many other approaches to quantization as well. Quantum theory may be viewed as a unitary irreducible representation (UIR) of the algebra of observables, the latter being selected by physical criteria [1]. The algebra itself must satisfy certain conditions so as to have the correct physical requirements. Generally it ends up as a C^*-algebra with further additional conditions equivalent to symmetries or other desirable properties (such as Lorentz invariance, relativistic causality) and so on. In relativistic field theory, this would lead to a von Neumann algebra. Here we are not going to pursue such an algebraic approach to quantization. Instead, we will consider the essential geometry (which has to do with the symplectic structure) of the classical theory and work out how a quantum theory can be constructed. This will be done in the language of Hamiltonians and Hilbert space. The key principle of quantization, as always, is that canonical transformations of the classical theory should be represented as unitary transformations on the Hilbert space of states in the quantum theory.

There is yet another approach to the quantum theory, the functional integral approach, which is formulated directly in terms of the action and can be made manifestly covariant if the theory of interest has relativistic invariance [2, 3]. Here we will not discuss this formulation either, but some points of overlap will be pointed out as the occasion arises.

Open Access This chapter is licensed under the terms of the Creative Commons Attribution 4.0 International License (http://creativecommons.org/licenses/by/4.0/), which permits use, sharing, adaptation, distribution and reproduction in any medium or format, as long as you give appropriate credit to the original author(s) and the source, provide a link to the Creative Commons license and indicate if changes were made.

The images or other third party material in this chapter are included in the chapter's Creative Commons license, unless indicated otherwise in a credit line to the material. If material is not included in the chapter's Creative Commons license and your intended use is not permitted by statutory regulation or exceeds the permitted use, you will need to obtain permission directly from the copyright holder.

Chapter 2
Symplectic Form and Poisson Brackets

We start with the formulation of theories in the symplectic language [4, 5]. Later, we will briefly discuss how this is connected to the action which may be used to specify the physical theory.

2.1 Symplectic Structure

In the analytical formulation of classical physics, the key concept is the phase space, which is a smooth even dimensional orientable manifold M (say, of dimension $2n$) endowed with a symplectic structure Ω. By this we mean that there is a differential 2-form Ω defined on M which is closed and nondegenerate. Closure means that $d\Omega = 0$, where d denotes exterior differentiation. The qualification "nondegenerate" refers to the fact that for any vector field ξ on M, if $i_\xi \Omega = 0$, then ξ must be zero. Here i_ξ indicates interior contraction with the vector field ξ. We will use q^μ to denote local coordinates on M. In terms of these, we can write

$$\Omega = \frac{1}{2}\, \Omega_{\mu\nu}\, dq^\mu \wedge dq^\nu \qquad (2.1)$$

The closure condition $d\Omega = 0$ can be written out as

$$
\begin{aligned}
d\Omega &\equiv \frac{1}{2}\frac{\partial \Omega_{\mu\nu}}{\partial q^\alpha}\, dq^\alpha \wedge dq^\mu \wedge dq^\nu \\
&= \frac{1}{3!}\left[\frac{\partial \Omega_{\mu\nu}}{\partial q^\alpha} + \frac{\partial \Omega_{\alpha\mu}}{\partial q^\nu} + \frac{\partial \Omega_{\nu\alpha}}{\partial q^\mu}\right] dq^\alpha \wedge dq^\mu \wedge dq^\nu \\
&= 0
\end{aligned}
\qquad (2.2)
$$

© The Author(s) 2024
V. P. Nair, *Geometric Quantization and Applications to Fields and Fluids*,
SpringerBriefs in Physics, https://doi.org/10.1007/978-3-031-65801-3_2

The contraction of Ω with a vector field $\xi = \xi^\mu (\partial/\partial q^\mu)$ is given by

$$i_\xi \Omega = \xi^\mu \Omega_{\mu\nu} \, dq^\nu, \qquad \xi = \xi^\mu \frac{\partial}{\partial q^\mu} \tag{2.3}$$

Thus in terms of components, the equation $i_\xi \Omega = 0$ becomes $\xi^\mu \Omega_{\mu\nu} = 0$. Nondegeneracy of Ω is then seen to be equivalent to the invertibility of $\Omega_{\mu\nu}$ as a matrix, so that $\xi^\mu \Omega_{\mu\nu} = 0$ implies $\xi^\mu = 0$; in other words, $\Omega_{\mu\nu}$, viewed as a matrix, does not have an eigenstate of eigenvalue equal to zero.

The inverse of $\Omega_{\mu\nu}$, which will be needed for some equations, will be denoted by $\Omega^{\mu\nu}$, with upper indices; i.e.,

$$\Omega_{\mu\nu} \, \Omega^{\nu\alpha} = \delta_\mu^{\ \alpha} \tag{2.4}$$

For now, we will take Ω to be nondegenerate. There are cases where the action will lead to a degenerate $\Omega_{\mu\nu}$; this occurs when the theory has a gauge symmetry. Elimination of certain components of the gauge field via gauge-fixing is then needed to define a nondegenerate Ω; we will consider such cases briefly later. With the structure Ω defined on it, M is a symplectic manifold.

Since Ω is closed, at least locally we can write

$$\Omega = d\mathcal{A} \tag{2.5}$$

The one-form $\mathcal{A}$ defined by this equation is called the canonical one-form or symplectic potential. There is an ambiguity in the definition of $\mathcal{A}$ since $\mathcal{A}$ and $\mathcal{A} + d\Lambda$ will give the same Ω for any function Λ on M. As we shall see shortly, this corresponds to the freedom of canonical transformations.

There are two types of features associated with the topology of the phase space which are apparent at this stage. The first question is: Is every 2-form Ω which is closed (i.e., obeys $d\Omega = 0$) the exterior derivative of a 1-form $\mathcal{A}$? In general, the answer is no. The set of linearly independent closed 2-forms which cannot be expressed as $d\mathcal{A}$ for some 1-form $\mathcal{A}$ is the second cohomology group of the manifold M; this is denoted by $\mathcal{H}^2(M)$.[1] Thus, if the phase space M has nontrivial second cohomology, i.e., if $\mathcal{H}^2(M) \neq 0$, then there are possible choices for Ω for which there is no globally defined potential $\mathcal{A}$. There are examples of physical interest where this happens. Such Ω's correspond to the Wess-Zumino terms in the action and are related to anomalies and also to central (and other) extensions of the algebra of observables.

Even when $\mathcal{H}^2(M) = 0$, there can be topological issues in defining $\mathcal{A}$. If the first cohomology $\mathcal{H}^1(M) \neq 0$, this means, by definition, that there are 1-forms A whose derivative is zero, but which are not of the form d of a function on M. Thus $\mathcal{A}$ and $\mathcal{A} + A$ will give the same Ω, but the difference is not just d for some function Λ, since A does not have to be of the form $d\Lambda$, globally. In other words, there are inequivalent

---

[1] We refer to the cohomologies over $\mathbb{R}$ since adding forms with real coefficients is what is appropriate.

$\mathcal{A}$'s for the same Ω. In these cases, one can consider the integral of $\mathcal{A}$ around closed noncontractible curves on M. The values of these integrals or holonomies will be important in the quantum theory as vacuum angles. The standard θ-vacuum of nonabelian gauge theories is an example. We will take up these topological issues in more detail later.

Given the symplectic structure, transformations which preserve Ω are evidently special; these are called *canonical transformations*. In other words, a canonical transformation is a diffeomorphism (or coordinate transformation) of M which preserves Ω. Infinitesimally, the coordinate transformation may be taken to be $q^\mu \to q^\mu + \xi^\mu(q)$. The change in Ω due to this is given by

$$
\begin{aligned}
\delta_\xi \, \Omega &= \frac{1}{2} \Omega_{\mu\nu}(q + \xi) \, \mathrm{d}(q^\mu + \xi^\mu) \wedge \mathrm{d}(q^\nu + \xi^\nu) - \frac{1}{2} \Omega_{\mu\nu} \, \mathrm{d}q^\mu \wedge \mathrm{d}q^\nu \\
&= \frac{1}{2} \left[\xi^\alpha \frac{\partial \Omega_{\mu\nu}}{\partial q^\alpha} + \Omega_{\alpha\nu} \frac{\partial \xi^\alpha}{\partial q^\mu} + \Omega_{\mu\alpha} \frac{\partial \xi^\alpha}{\partial q^\nu} \right] \mathrm{d}q^\mu \wedge \mathrm{d}q^\nu \\
&= \frac{1}{2} \xi^\alpha \left[\frac{\partial \Omega_{\mu\nu}}{\partial q^\alpha} + \frac{\partial \Omega_{\alpha\mu}}{\partial q^\nu} + \frac{\partial \Omega_{\nu\alpha}}{\partial q^\mu} \right] \mathrm{d}q^\mu \wedge \mathrm{d}q^\nu \\
&\quad + \frac{1}{2} \left[\partial_\mu(\xi^\alpha \Omega_{\alpha\nu}) - \partial_\nu(\xi^\alpha \Omega_{\alpha\mu}) \right] \mathrm{d}q^\mu \wedge \mathrm{d}q^\nu \\
&= i_\xi(\mathrm{d}\Omega) + \mathrm{d}(i_\xi \Omega)
\end{aligned}
\tag{2.6}
$$

For a canonical transformation, this change must be zero.[2] Since $\mathrm{d}\Omega = 0$, this means that canonical transformations are generated by vector fields ξ such that

$$
\mathrm{d}\,(i_\xi \Omega) = 0
\tag{2.7}
$$

Thus for canonical transformations, $i_\xi \Omega$ is a closed 1-form. If the first cohomology $\mathcal{H}^1(M)$ of M is trivial, we can write

$$
i_\xi \Omega = -\mathrm{d}f, \qquad \xi^\alpha \Omega_{\alpha\nu} = -\frac{\partial f}{\partial q^\nu}
\tag{2.8}
$$

for some function f on M. *In other words, to every infinitesimal canonical transformation, we can associate a function on M.*[3] Since Ω is invertible, we can always associate a vector field to a function f by the correspondence

$$
\xi^\mu = \Omega^{\mu\nu} \partial_\nu f
\tag{2.9}
$$

---

[2] The right hand side of (2.6) is the Lie derivative of Ω with respect to the vector field $\xi^\alpha (\partial/\partial q^\alpha)$.

[3] If $\mathcal{H}^1(M) \neq 0$, then there is the possibility that for some transformations ξ, the corresponding $i_\xi \Omega$ in a nontrivial element of $\mathcal{H}^1(M)$ and hence there is no globally defined function f for this transformation. As mentioned before this is related to the possibility of vacuum angles in the quantum theory. For the moment, we shall consider the case $\mathcal{H}^1(M) = 0$.

What we are saying now is that, for a canonical transformation, we can go the other way, associating a function with the vector field which gives the canonical transformation, at least when $\mathcal{H}^1(M) = 0$. There is a one-to-one mapping between functions on M and vector fields corresponding to infinitesimal canonical transformations. A vector field corresponding to an infinitesimal canonical transformation is often referred to as a Hamiltonian vector field. The function f defined by (2.8) is called the generating function for the canonical transformation corresponding to the vector field.

It is important that for every function f on the phase space M, we can associate a Hamiltonian vector field as in (2.11). This means that all observables which are functions on M generate canonical transformations.

2.2 Poisson Brackets

Let ξ, η be two Hamiltonian vector fields which means that they preserve Ω; let their generating functions be f and g respectively. The Lie bracket or commutator of ξ and η is given in local coordinates by

$$[\xi, \eta]^\mu = \xi^\nu \partial_\nu \eta^\mu - \eta^\nu \partial_\nu \xi^\mu \tag{2.10}$$

We can easily verify that the commutator will also preserve Ω. We must therefore have a function corresponding to $[\xi, \eta]$. It is designated as the Poisson bracket of g and f and is denoted by $\{g, f\}$. Explicitly, we define the Poisson bracket (PB) as

$$\begin{aligned}
\{f, g\} &= i_\xi i_\eta \Omega = \eta^\mu \xi^\nu \Omega_{\mu\nu} \\
&= -i_\xi dg = i_\eta df \\
&= \Omega^{\mu\nu} \partial_\mu f \partial_\nu g
\end{aligned} \tag{2.11}$$

Notice that, for the choice $f = q^\mu$, $g = q^\nu$, this reduces to

$$\{q^\mu, q^\nu\} = \Omega^{\mu\nu} \tag{2.12}$$

Because of the antisymmetry of $\Omega_{\mu\nu}$, the Poisson bracket has the property

$$\{f, g\} = -\{g, f\} \tag{2.13}$$

Further, from the definition, we can write, using local coordinates,

$$
\begin{aligned}
2\,\partial_\alpha\{f,g\} &= \partial_\alpha(\eta\cdot\partial f - \xi\cdot\partial g)\\
&= \partial_\alpha\eta^\mu\partial_\mu f + \eta^\mu(\partial_\mu\partial_\alpha f) - \partial_\alpha\xi^\mu\partial_\mu g - \xi^\mu(\partial_\mu\partial_\alpha g)\\
&= \partial_\alpha\eta^\mu\partial_\mu f - \partial_\alpha\xi^\mu\partial_\mu g + \eta\cdot\partial(\xi^\mu\Omega_{\alpha\mu}) - \xi\cdot\partial(\eta^\mu\Omega_{\alpha\mu})\\
&= \partial_\alpha\eta^\mu\partial_\mu f - \partial_\alpha\xi^\mu\partial_\mu g + (\xi\cdot\partial\eta - \eta\cdot\partial\xi)^\mu\Omega_{\mu\alpha}\\
&\quad + \eta^\mu\xi^\nu(\partial_\mu\Omega_{\alpha\nu} + \partial_\nu\Omega_{\mu\alpha})\\
&= [\xi,\eta]^\mu\Omega_{\mu\alpha} + \partial_\alpha(\eta^\mu\xi^\nu\Omega_{\mu\nu})\\
&\quad + \eta^\mu\xi^\nu(\partial_\mu\Omega_{\alpha\nu} + \partial_\nu\Omega_{\mu\alpha} + \partial_\alpha\Omega_{\nu\mu})
\end{aligned}
\tag{2.14}
$$

In local coordinates, the closure of Ω is the statement $\partial_\mu\Omega_{\alpha\nu} + \partial_\nu\Omega_{\mu\alpha} + \partial_\alpha\Omega_{\nu\mu} = 0$. From the equation given above, we then see that

$$
-\,\mathrm{d}\{g,f\} = i_{[\xi,\eta]}\,\Omega
\tag{2.15}
$$

This result shows the correspondence stated earlier.

Consider now the change in a function F due to a canonical transformation $q^\mu \to q^\mu + \xi^\mu$. The change in F is obviously $\xi^\mu\partial_\mu F$. Let f be the function corresponding to the vector field ξ^μ via the correspondence (2.8). We can write the change in F as

$$
\delta F = \xi^\mu\partial_\mu F = (\Omega^{\mu\alpha}\partial_\alpha f)\,\partial_\mu F = \{F,f\}
\tag{2.16}
$$

Thus the change in a function F due to the canonical transformation $q^\mu \to q^\mu + \xi^\mu$ is given by the Poisson bracket of F with the generating function f corresponding to the vector field ξ.

Another important property of the Poisson bracket is the Jacobi identity. For any three functions f, g, h, we have the identity

$$
\{f,\{g,h\}\} + \{h,\{f,g\}\} + \{g,\{h,f\}\} = 0
\tag{2.17}
$$

This can be verified by direct computation from the definition of the Poisson bracket. In fact, if ξ, η, ρ are the Hamiltonian vector fields corresponding to the functions f, g, h, then, by direct computation,

$$
\{f,\{g,h\}\} + \{h,\{f,g\}\} + \{g,\{h,f\}\} = i_\xi i_\eta i_\rho(\mathrm{d}\Omega)
\tag{2.18}
$$

and so the Jacobi identity (2.17) follows from the closure of Ω.

An expression which will be useful later is the change of the symplectic potential $\mathcal{A}$ under an infinitesimal canonical transformation; this can be worked out as

$$\delta_\xi \mathcal{A} = \mathcal{A}_\mu(q + \xi)d(q^\mu + \xi^\mu) - \mathcal{A}_\mu(q)dq^\mu = \left[\xi^\alpha \partial_\alpha \mathcal{A}_\mu + \mathcal{A}_\alpha \frac{\partial \xi^\alpha}{\partial q^\mu}\right] dq^\mu$$

$$= \left[\xi^\alpha(\partial_\alpha \mathcal{A}_\mu - \partial_\mu \mathcal{A}_\alpha) + \partial_\mu(\xi^\alpha \mathcal{A}_\alpha)\right] dq^\mu$$

$$= \left[\xi^\alpha \Omega_{\alpha\mu} + \partial_\mu(\xi^\alpha \mathcal{A}_\alpha)\right] dq^\mu$$

$$= \partial_\mu (\xi^\alpha \mathcal{A}_\alpha - f) \, dq^\mu \tag{2.19}$$

where we used the definition of Ω and the fact that ξ is a Hamiltonian vector field with a corresponding function f defined by Eq. (2.8). Equation (2.19) shows that under a canonical transformation $\mathcal{A} \to \mathcal{A} + d\Lambda$, $\Lambda = i_\xi \mathcal{A} - f$. Evidently $d\mathcal{A} = \Omega$ is unchanged under such a transformation. This suggests a very useful way of thinking about these structures.

> We may view $\mathcal{A}$ as a $U(1)$ gauge potential and Ω as the corresponding field strength. The transformation $\mathcal{A} \to \mathcal{A} + d\Lambda$ is thus a gauge transformation. We can use this point of view to construct an invariant description, using covariant derivatives and other properly transforming quantities.

A remark comparing the symplectic language we have used to some other formulations may be useful at this point. If we use the definition of the Poisson bracket, namely (2.11), for the phase space coordinates themselves, we have Eq. (2.12), $\{q^\mu, q^\nu\} = \Omega^{\mu\nu}$. This is often interpreted as saying that the "basic Poisson brackets" (i.e., PBs for the phase space coordinates themselves) are the inverse of the symplectic structure.

2.3 Phase Volume

The symplectic two-form can be used to define a volume form on the phase space M by

$$d\sigma(M) = c \, \frac{\Omega \wedge \Omega \wedge \cdots \wedge \Omega}{(2\pi)^n} = c \, n! \sqrt{\det\left(\frac{\Omega}{2\pi}\right)} \, d^{2n}q \tag{2.20}$$

where we take the n-fold product of Ω's for a $2n$-dimensional phase space and the second expression involves the determinant of $\Omega_{\mu\nu}$ as a matrix. (c is a constant which is undetermined at this stage.) If the dimension of the phase space is infinite, then a suitable regularized form of the determinant has to be used, with $n \to \infty$ at the end. The volume measure defined by Eq. (2.20) is called the Liouville measure. Since it is defined in terms of Ω, this volume element is invariant under canonical transformations.

2.4 Darboux's Theorem

A useful result concerning the symplectic form is Darboux's theorem which states that in the neighbourhood of a point on the phase space it is possible to choose coordinates $p_i, x^i, i = 1, 2, \ldots, n$, (which are functions of the coordinates q^μ we started with) such that the symplectic two-form is

$$\Omega = \mathrm{d}p_i \wedge \mathrm{d}x^i = \frac{1}{2} J_{\mu\nu} \, \mathrm{d}Q^\mu \wedge \mathrm{d}Q^\nu$$

$$Q^\mu = (p_1, x^1, p_2, x^2, \ldots, p_n, x^n) \tag{2.21}$$

The tensor $J_{\mu\nu}$ (which is $\Omega_{\mu\nu}$ in this coordinate system) can be expressed in matrix form as

$$J_{\mu\nu} = \begin{bmatrix} 0 & 1 & 0 & 0 & \cdots \\ -1 & 0 & 0 & 0 & \cdots \\ 0 & 0 & 0 & 1 & \cdots \\ 0 & 0 & -1 & 0 & \cdots \\ \cdots & \cdots & \cdots & \cdots & \cdots \end{bmatrix} \tag{2.22}$$

Evidently from the form of Ω, we see that the Poisson brackets in terms of this set of coordinates are

$$\{x^i, x^j\} = 0$$
$$\{x^i, p_j\} = \delta^i{}_j$$
$$\{p_i, p_j\} = 0 \tag{2.23}$$

An elegant proof of this theorem can be found in Arnold's book [4]. Here we will give a short rephrasing of the same argument. The simplest way to prove the theorem is by induction. First of all, we note that the Darboux theorem is equivalent to the statement that one can choose coordinates such that the fundamental Poisson brackets are given by (2.23).

We now start with a point P in some neighborhood of the manifold M. Our attempt is to reduce Ω to the Darboux form in this neighborhood. We start by taking any nonconstant function of the coordinates q^α as the first coordinate p_1, with $\mathrm{d}p_1 \neq 0$ at P. We can also assume $p_1 = 0$ at P, since this can always be done by adding a constant to any chosen p_1. Since p_1 is a function in the neighborhood, there is a Hamiltonian vector field

$$P_1 = \Omega^{\alpha\beta} \frac{\partial p_1}{\partial q^\beta} \tag{2.24}$$

Consider now a set of trajectories defined by

$$\frac{d\xi^\alpha}{d\tau} = -\Omega^{\alpha\beta} \frac{\partial p_1}{\partial q^\beta} \tag{2.25}$$

These are flow lines generated by the vector field P_1, with ξ^α as the values of the coordinates q^α on the trajectories, as functions of τ. Now choose a $(2n - 1)$-dimensional surface Σ which intersects these flow lines transversally and contains the point P.

Consider any point q (with coordinates q^α) near Σ but not necessarily on it. We can solve (2.25) with q as the initial point and choose the direction such that the motion is towards the surface Σ. At some value τ determined by the initial point q, this motion arrives at Σ. We denote this particular value of τ, viewed as a function of the coordinates q^α of the starting point q, by x_1. If τ is taken to be infinitesimal, this is given by

$$\xi^\alpha(\tau)\big|_\Sigma = q^\alpha - \Omega^{\alpha\beta}\frac{\partial p_i}{\partial q^\beta}\tau(q) \tag{2.26}$$

Equivalently, we can write this as

$$\Omega^{\alpha\beta}\frac{\partial p_i}{\partial q^\beta}\tau(q) = q^\alpha - \xi^\alpha(\tau)\big|_\Sigma \tag{2.27}$$

This equation shows that for this function $x_1(q) = \tau(q)$ we have

$$\{x_1, p_1\} = 1 \tag{2.28}$$

We can thus take p_1, x_1 as the first pair of Darboux coordinates. Notice that $x_1 = 0$ for points on Σ.

We now consider a subspace, a surface Σ^* defined by $p_1 = 0, x_1 = 0$. The differentials dp_1, dx_1 are linearly independent since the Poisson bracket of x_1 and p_1 is nonzero. Thus they define two noncollinear directions which take us off the surface. Therefore the surface Σ^* is $(2n - 2)$-dimensional. If X_1 and P_1 denote the vector fields corresponding to x_1 and p_1 respectively, we have $i_{X_1}\Omega = -dx_1, i_{P_1}\Omega = -dp_1$. This result, along with (2.28) shows that we can write

$$\Omega = \Omega^* + dp_1 \wedge dx_1 \tag{2.29}$$

where Ω^* does not involve differentials dp_1 or dx_1. (If it did, we would have a contradiction with the contraction of X_1 and P_1 with Ω in comparison to (2.28).) Now consider a vector field χ which generates a flow along (i.e. tangential to) Σ^*. By definition, χ cannot change the value of p_1, x_1, so we have $i_\chi dp_1 = 0, i_\chi dx_1 = 0$. This means that the contraction of any vector tangential to Σ^* with Ω^* is the same as its contraction with Ω. Thus Ω^* must be invertible for vectors tangential to Σ^*. Finally, it is evident that $d\Omega^* = 0$. Ω^* therefore defines a symplectic two-form on the $(2n - 2)$-dimensional subspace Σ^*. The problem has thus been reduced to the question of the existence of the Darboux coordinates on the lower dimensional space. We can now proceed in similar manner, starting with Σ^* and Ω^*. We can construct another canonical pair p_2, x_2 and reduce the problem to a $(2n - 4)$-dimensional subspace, and so on inductively to complete the proof of the theorem.

Problem

2.1 Derive Eq. (2.18) from the definition of Poisson brackets.

Open Access This chapter is licensed under the terms of the Creative Commons Attribution 4.0 International License (http://creativecommons.org/licenses/by/4.0/), which permits use, sharing, adaptation, distribution and reproduction in any medium or format, as long as you give appropriate credit to the original author(s) and the source, provide a link to the Creative Commons license and indicate if changes were made.

The images or other third party material in this chapter are included in the chapter's Creative Commons license, unless indicated otherwise in a credit line to the material. If material is not included in the chapter's Creative Commons license and your intended use is not permitted by statutory regulation or exceeds the permitted use, you will need to obtain permission directly from the copyright holder.

Chapter 3
Classical Dynamics

The importance of the symplectic approach is that, classically, the time-evolution of any quantity is a particular canonical transformation generated by a function H called the Hamiltonian. This is the essence of the Hamiltonian formulation of dynamics. Thus if F is any function on M, we then have

$$\frac{\partial F}{\partial t} = \{F, H\} \tag{3.1}$$

Specifically for the local coordinates q^μ on M this equation leads to

$$\frac{\partial q^\mu}{\partial t} = \{q^\mu, H\} = \Omega^{\mu\nu}\frac{\partial H}{\partial q^\nu} \tag{3.2}$$

Since Ω is invertible, we can also write this equation as

$$\Omega_{\mu\nu}\frac{\partial q^\nu}{\partial t} = \frac{\partial H}{\partial q^\mu} \tag{3.3}$$

If we use the Darboux coordinates (p_i, x^i), these equations (either (3.2) or (3.3)) become

$$\dot{p}_i = -\frac{\partial H}{\partial x^i}, \qquad \dot{x}^i = \frac{\partial H}{\partial p_i} \tag{3.4}$$

which are more easily recognizable as Hamilton's canonical equations.

We are now in a position to connect the dynamics to an action and a variational principle. We *define* the action as

$$S = \int_{t_i}^{t_f} dt \left(A_\mu \frac{dq^\mu}{dt} - H \right) \tag{3.5}$$

© The Author(s) 2024
V. P. Nair, *Geometric Quantization and Applications to Fields and Fluids*,
SpringerBriefs in Physics, https://doi.org/10.1007/978-3-031-65801-3_3

where $q^\mu(t)$ gives a path on M. Under a general variation of the path $q^\mu(t) \to q^\mu(t) + \xi^\mu(t)$, the action changes by

$$\delta S = \int dt \left(\frac{\partial \mathcal{A}_\nu}{\partial q^\mu} \frac{dq^\nu}{dt} \xi^\mu + \mathcal{A}_\mu \frac{d\xi^\mu}{dt} - \frac{\partial H}{\partial q^\mu} \xi^\mu \right)$$

$$= \mathcal{A}_\mu \xi^\mu \Big]_{t_i}^{t_f} + \int dt \left(\Omega_{\mu\nu} \frac{dq^\nu}{dt} - \frac{\partial H}{\partial q^\mu} \right) \xi^\mu \tag{3.6}$$

The variational principle says that the equations of motion are given by the extremization of the action, i.e., by $\delta S = 0$, for the restricted set of variations with the boundary data (initial and final end point data) fixed. From the above variation, we see that this gives the Hamiltonian equations of motion (3.3). There is a slight catch in this argument because q^μ are phase space coordinates and obey first order equations of motion. So we can only specify the initial value of q^μ. However, the Darboux theorem tells us that one can choose coordinates on M such that the canonical one-form $\mathcal{A}$ is of the form $p_i dx^i$. As a result, the ξ^μ in the boundary term is just δx^i. Therefore, instead of specifying initial data for all q^μ, we can choose to specify initial and final data for the x^i's. Since the boundary values are to be kept fixed in the variational principle $\delta S = 0$, we may set $\delta x^i = 0$ at both boundaries and the equations of motion are indeed just (3.3).

We have shown how to define the action if Ω is given. However, going back to the general variations, notice that the boundary term resulting from the time-integration is just the canonical one-form contracted with ξ^μ. Thus if we start from the action as the given quantity, we can identify the canonical one-form and hence Ω from the boundary term which arises in a general variation. In fact

$$\delta S = i_\xi \mathcal{A}(t_f) - i_\xi \mathcal{A}(t_i) + \int dt \left(\Omega_{\mu\nu} \frac{dq^\nu}{dt} - \frac{\partial H}{\partial q^\mu} \right) \xi^\mu \tag{3.7}$$

As an example of this, consider a real scalar field theory with the action

$$S = \int d^4x \left[\frac{1}{2} \dot\varphi^2 - \frac{1}{2} (\nabla\varphi)^2 - \frac{1}{2} m^2\varphi^2 - \alpha\varphi^4 \right] \tag{3.8}$$

The variation of the action leads, upon time-integration, to the boundary term

$$\delta S = \int d^3x \, \dot\varphi \, \delta\varphi \Big]_{t_i}^{t_f} + \int d^4x \, [\cdots] \tag{3.9}$$

The canonical 1-form or the symplectic potential (at a fixed time t) can thus be taken as

$$\mathcal{A} = \int d^3x \, \dot\varphi \, \delta\varphi \tag{3.10}$$

In this analysis, we are at a fixed time, so $\dot{\varphi}$ is a function independent of φ. The phase space thus consists of the set of functions $\{\dot{\varphi}, \varphi\}$ on the three-dimensional space $\mathbb{R}^3$. $\mathcal{A}$ in (3.10) is a 1-form on the phase space, if we interpret $\delta\varphi$ (which is a functional variation) as the exterior derivative on the space of fields.[1]

If we add a total derivative to the Lagrangian, say, $\mathcal{S} \to \mathcal{S} + \int dt\, \dot{f}$, it does not affect the equations of motion. However, the new $\mathcal{A}$ obtained from the boundary values has an extra term δf. This is the exterior derivative of f and hence the symplectic two-form Ω (which is $\delta\mathcal{A}$) is unchanged. We see that the freedom of adding total derivatives to the Lagrangian is thus the freedom of canonical transformations.

An interesting variant for the scalar field theory is to consider the light-cone quantization of the same theory. Introduce light-cone coordinates, corresponding to a light-cone in the z-direction, as

$$u = \frac{1}{\sqrt{2}}(t + z), \qquad v = \frac{1}{\sqrt{2}}(t - z) \tag{3.11}$$

Instead of considering evolution of the fields in time t, we can consider evolution in one of the light-cone coordinates, say, u. The analog of 'space' is given by the other light-cone coordinate v and the two coordinates $x^{\mathrm{T}} = x, y$ transverse to the light-cone. They correspond to equal-u hypersurfaces. The action (3.8) for the real scalar field $\varphi(u, v, x, y)$ can be written in these coordinates as

$$\mathcal{S} = \int du\, dv\, d^2x^{\mathrm{T}} \left[\partial_u\varphi\partial_v\varphi - \tfrac{1}{2}(\partial_{\mathrm{T}}\varphi)^2 - \tfrac{1}{2}m^2\varphi^2 - \alpha\varphi^4 \right] \tag{3.12}$$

This is of the first order in the u-derivatives, which are the analog of the time-derivatives. The time-integration of the variation of this action leads to the boundary term

$$\delta\mathcal{S} = \int dv\, d^2x^{\mathrm{T}}\, \partial_v\varphi\, \delta\varphi \Big]_{u_i}^{u_f} + \text{volume integral} \tag{3.13}$$

Since $\partial_v\varphi$ is a spatial derivative now, it is not independent of φ and so the phase space is given by field configurations $\varphi(v, x^{\mathrm{T}})$. The symplectic potential is

$$\mathcal{A} = \int dv\, d^2x^{\mathrm{T}}\, \partial_v\varphi\, \delta\varphi \tag{3.14}$$

Taking the exterior derivative (denoted by the symbol δ on the space of fields), we find the symplectic two-form as[2]

---

[1] In the context of field theory we will often use δ to indicate the exterior derivative on the space of fields.

[2] There is a wedge product for the two differentials in this equation. To avoid clutter, we will not write the wedge symbol from now on if it is clear from the context.

$$\Omega = \int dv\, d^2x^{\mathrm{T}}\, (\partial_v \delta\varphi)\, \delta\varphi \tag{3.15}$$

The contraction of this with the vector field $\xi = \delta/\delta\varphi$ gives

$$i_\xi \Omega = -2\partial_v(\delta\varphi) = -\delta(2\partial_v\varphi) \tag{3.16}$$

The function corresponding to the vector field ξ is thus $2\,\partial_v\varphi$. Using (2.11), we then find

$$\{2\partial_v\varphi(v, x^{\mathrm{T}}), \varphi(v', x'^{\mathrm{T}})\} = -i_\xi \delta\varphi(v', x'^{\mathrm{T}}) = -\delta(v - v')\delta(x^{\mathrm{T}} - x'^{\mathrm{T}}) \tag{3.17}$$

Keeping in mind the antisymmetry of the Poisson brackets, this is equivalent to

$$\{\varphi(v, x^{\mathrm{T}}), \varphi(v', x'^{\mathrm{T}})\} = -\frac{1}{4}\epsilon(v - v')\delta(x^{\mathrm{T}} - x'^{\mathrm{T}}) \tag{3.18}$$

where $\epsilon(v - v')$ is the signature function given by

$$\epsilon(v - v') = \begin{cases} 1 & v > v' \\ -1 & v < v' \end{cases} \tag{3.19}$$

By using the symplectic form, we are directly led to the Poisson bracket (3.18). In the more conventional approach where we consider the canonical momentum Π, the present situation has a constraint $\Pi - \partial_v\varphi \approx 0$. In such a formulation, one has to use Dirac's theory of constraints to obtain the basic Poisson bracket. The use of Ω bypasses these steps.

We will consider other cases of determining the symplectic form using this method (of identifying the surface term from the time-integration in the variation of the action) when we take up examples.

Open Access This chapter is licensed under the terms of the Creative Commons Attribution 4.0 International License (http://creativecommons.org/licenses/by/4.0/), which permits use, sharing, adaptation, distribution and reproduction in any medium or format, as long as you give appropriate credit to the original author(s) and the source, provide a link to the Creative Commons license and indicate if changes were made.

The images or other third party material in this chapter are included in the chapter's Creative Commons license, unless indicated otherwise in a credit line to the material. If material is not included in the chapter's Creative Commons license and your intended use is not permitted by statutory regulation or exceeds the permitted use, you will need to obtain permission directly from the copyright holder.

Chapter 4
Geometric Quantization

Quantum theory of any physical system is a unitary irreducible representation of the algebra of observables of the system. This means that the observables are realized as linear operators on a Hilbert space. The allowed transformation of variables are then unitary transformations. There are thus two key points regarding quantization:

1. We need a correspondence between canonical transformations and unitary transformations.
2. We must ensure that the representation of unitary transformations on the Hilbert space is irreducible.

Since functions on phase space generate canonical transformations and hermitian operators generate unitary transformations, the first point ensures that we get a correspondence between functions on phase space and operators on the Hilbert space. The algebra of Poisson brackets will be replaced by the algebra of commutation rules for the operators. The irreducibility leads to the necessity of choosing a polarization for the wave functions. Some general references on geometric quantization are [5, 7, 8].

4.1 Pre-Quantization

We will first consider the notion of the wave function before discussing how operators act on such wave functions. In the geometric approach, the first step is the so-called prequantum line bundle.

This is a complex line bundle on the phase space with curvature Ω. Sections of this line bundle form the prequantum Hilbert space. In less technical terms, we utilize the similarity we mentioned earlier, namely, that the symplectic potential may be thought of as a $U(1)$ gauge field, with the transformations $\mathcal{A} \to \mathcal{A} + d\Lambda$ viewed as a gauge transformation. We can then consider complex functions $\Psi(q)$ defined on

© The Author(s) 2024
V. P. Nair, *Geometric Quantization and Applications to Fields and Fluids*,
SpringerBriefs in Physics, https://doi.org/10.1007/978-3-031-65801-3_4

open neighborhoods in M. These are like matter fields, they are the sections of the line bundle. This means that locally they are complex functions which transform as

$$\Psi \rightarrow \Psi' = \exp(\mathrm{i}\,\Lambda)\,\Psi \tag{4.1}$$

We can define a covariant derivative acting on $\Psi(q)$ using $\mathcal{A}$ as

$$\mathcal{D}_\mu \Psi \equiv \left(\frac{\partial}{\partial q^\mu} - \mathrm{i}\,\mathcal{A}_\mu\right)\Psi \tag{4.2}$$

The commutator of two covariant derivatives gives $-\mathrm{i}\,\Omega$, this is the meaning of saying that the curvature of the line bundle is Ω.

Since canonical transformations correspond to $\mathcal{A} \rightarrow \mathcal{A} + \mathrm{d}\Lambda$, the transformation of Ψ as given in (4.1) is equivalent to the requirement of canonical transformations being implemented as unitary transformations. The transition rules for the Ψ's from one patch on M to another are likewise given by exponentiating the transition function for $\mathcal{A}$. The functions Ψ's so defined form the prequantum Hilbert space with the inner product

$$(1|2) = \int \mathrm{d}\sigma(M)\,\Psi_1^*\,\Psi_2 \tag{4.3}$$

where $\mathrm{d}\sigma(M)$ is the Liouville measure (2.20) on the phase space defined by Ω.

The next step will be to define operators (acting on Ψ) corresponding to various functions on the phase space. A function $f(q)$ on the phase space generates a canonical transformation which leads to the change $\Lambda = i_\xi \mathcal{A} - f$ in the symplectic potential, see (2.19). The corresponding change in Ψ is thus

$$\begin{aligned}
\delta\Psi &= \xi^\mu \partial_\mu \Psi - \mathrm{i}(i_\xi \mathcal{A} - f)\Psi \\
&= \xi^\mu\left(\partial_\mu - \mathrm{i}\mathcal{A}_\mu\right)\Psi + \mathrm{i}f\Psi \\
&= \left(\xi^\mu \mathcal{D}_\mu + \mathrm{i}f\right)\Psi
\end{aligned} \tag{4.4}$$

where the first term on the right hand side in the first line gives the change in Ψ considered as a function and the second term compensates for the change of $\mathcal{A}$. The change can be expressed using the covariant derivative as in the last line. Given (4.4), it is natural to define the *prequantum operator* corresponding to $f(q)$ by

$$\mathcal{P}(f) = -\mathrm{i}\left(\xi \cdot \mathcal{D} + \mathrm{i}f\right) \tag{4.5}$$

We can easily check that

$$\int \mathrm{d}^{2n}q\,\sqrt{\det \Omega}\,\Psi_1^*\left[\mathcal{P}(f)\,\Psi_2\right] = \int \mathrm{d}^{2n}q\,\sqrt{\det \Omega}\,\left[\mathcal{P}(f)\Psi_1\right]^*\,\Psi_2 \tag{4.6}$$

so that $\mathcal{P}(f)$ is a symmetric operator, which is a necessary condition for a unitary representation. (Strictly speaking, before we can claim a unitary representation, we need to consider the completion of the set of such functions and also make sure the domains and ranges of operators match; we will not go into this question, since the whole issue has to be addressed for the true wave functions anyway.)

Now consider the algebra of the prequantum operators. We have already seen in (2.15) that if the Hamiltonian vector fields for f, g are ξ and η respectively, then the vector field corresponding to the Poisson bracket $\{f, g\}$ is $-[\xi, \eta]$. Using the definition of the prequantum operator above, we then find

$$
\begin{aligned}
[\mathcal{P}(f), \mathcal{P}(g)] &= [-i\xi \cdot \mathcal{D} + f, -i\eta \cdot \mathcal{D} + g] \\
&= -\left[\xi^\mu \mathcal{D}_\mu, \eta^\nu \mathcal{D}_\nu\right] - i\xi^\mu[\mathcal{D}_\mu, g] + i\eta^\mu[\mathcal{D}_\mu, f] \\
&= i\xi^\mu \eta^\nu \Omega_{\mu\nu} - (\xi^\mu \partial_\mu \eta^\nu)\mathcal{D}_\nu + (\eta^\mu \partial_\mu \xi^\nu)\mathcal{D}_\nu - i\xi^\mu \partial_\mu g + i\eta^\mu \partial_\mu f \\
&= i\left(-\xi^\mu \eta^\nu \Omega_{\mu\nu} + i[\xi, \eta] \cdot \mathcal{D}\right) \\
&= i\left(-i\left(i_{[\eta,\xi]}\mathcal{D}\right) + \{f, g\}\right) \\
&= i\,\mathcal{P}(\{f, g\})
\end{aligned}
\tag{4.7}
$$

In other words, the prequantum operators form a representation of the Poisson bracket algebra of functions on phase space.

4.2 Polarization

It seems like we have all the ingredients for the quantum theory, but not quite so. The prequantum wave functions Ψ depend on all phase space variables. The representation of the Poisson bracket algebra on such wave functions, given by the prequantum operators, is reducible. A simple example will suffice to illustrate this point.

Consider a point particle in one dimension, with the symplectic two-form $\Omega = dp \wedge dx$. We can choose $\mathcal{A} = p\,dx$. The vector fields corresponding to x and p are $\xi_x = -\partial/\partial p$ and $\xi_p = \partial/\partial x$. The corresponding prequantum operators are

$$
\mathcal{P}(x) = i\frac{\partial}{\partial p} + x, \qquad \mathcal{P}(p) = -i\frac{\partial}{\partial x}
\tag{4.8}
$$

which obey the commutation rule

$$
[\mathcal{P}(x), \mathcal{P}(p)] = i
\tag{4.9}
$$

We have a representation of the algebra of $\mathcal{P}(x)$, $\mathcal{P}(p)$ in terms of the prequantum wave functions $\Psi(x, p)$. But this is reducible. For if we consider the subset of functions on the phase space which are independent of p, namely those which obey the condition

$$\frac{\partial \Psi}{\partial p} = 0, \tag{4.10}$$

then the prequantum operators reduce to

$$\mathcal{P}(x) = x, \qquad \mathcal{P}(p) = -i\frac{\partial}{\partial x} \tag{4.11}$$

which obey the same algebra (4.9). Thus we are able to obtain a representation of the algebra of observables on the smaller space of Ψ's obeying the constraint (4.10), showing that the previous representation (4.8) is reducible.

In order to obtain an irreducible representation, one has to impose subsidiary conditions which restrict the dependence of the prequantum wave functions to half the number of phase space variables. This is the choice of polarization and generally leads to an irreducible representation of the Poisson algebra.

If we are talking about ordinary functions f on the phase space M, the statement that f is independent of n of the coordinates can be phrased as

$$P_i^\mu \frac{\partial f}{\partial q^\mu} = 0 \tag{4.12}$$

where $P_i = P_i^\mu(\partial/\partial q^\mu)$, $i = 1, 2, \ldots, n$, form n linearly independent vector fields. An integrability requirement for (4.12) is

$$[P_i, P_j]^\mu \frac{\partial f}{\partial q^\mu} = 0 \tag{4.13}$$

which can be ensured if

$$[P_i, P_j] = C_{ij}^k P_k \tag{4.14}$$

where the coefficients C_{ij}^k need not be constants. If we have a set of vector fields P_i obeying (4.14), then they are said to be in involution. If this is satisfied, we can integrate, starting from some point on M, along these vector fields and obtain, at least locally, a neighborhood of an n-dimensional submanifold. (This is ensured by Frobenius' theorem.) Such a submanifold is said to be a Lagrangian submanifold if we also have the condition

$$\Omega_{\mu\nu} P_i^\mu P_j^\nu = 0 \tag{4.15}$$

The prequantum wave functions are not functions on M, they are sections of a line bundle, i.e., they transform with a phase under $\mathcal{A} \to \mathcal{A} + d\Lambda$, and so we must impose the covariant version of (4.12). Thus, as the polarization condition we choose

$$P_i^\mu \mathcal{D}_\mu \Psi = 0 \tag{4.16}$$

where P_i are n linearly independent vector fields obeying (4.14) and (4.15). The integrability requirement for (4.16) is automatically satisfied since

$$[P_i^\mu \mathcal{D}_\mu, P_j^\nu \mathcal{D}_\nu] \Psi = C_{ij}^k P_k^\mu \mathcal{D}_\mu \Psi - \mathrm{i}\, \Omega_{\mu\nu} P_i^\mu P_j^\nu \Psi$$
$$= 0 \tag{4.17}$$

by virtue of (4.15) and (4.16). The prequantum wave functions restricted by the polarization condition (4.16) are the true wave functions of the theory. There can be different possible choices for the polarization leading to wave functions depending on different subsets of phase space coordinates. For example, the difference between the momentum space wave functions and the coordinate space wave functions familiar from elementary quantum mechanics is one of different polarization choices.

4.3 Measure of Integration

The next step is to define an inner product to make these wave functions into a Hilbert space. Obviously, if the wave functions do not depend on half the number of phase space coordinates, it does not make sense to integrate over them in an inner product. In particular, it would give an undefined or infinite value if those directions do not have a finite volume. So one needs to define a volume measure for integration over those directions or coordinates on which the wave functions do have a dependence. The problem is that while the Liouville measure for all of phase space is naturally defined in terms of the symplectic structure, there is no natural choice of integration measure for the reduced set of variables, once we impose the polarization requirement. In many cases, the phase space is the cotangent bundle of some manifold (which is the configuration space Q), which means that it is made of the coordinates and co-vectors. For example, for particle dynamics on $Q = \mathbb{R}^3$, $M = T^*\mathbb{R}^3$. The usual coordinates x^μ and the momenta p_μ are the basic coordinates for M. Then, if we use a polarization given by $P_i^\mu = (\partial/\partial p_\mu)$, the wave functions depend on x^μ only. This is the usual coordinate space Schrödinger quantum mechanics and one can use the integration just on $\mathbb{R}^3$ to form the inner product. But generally speaking, unless M is the cotangent bundle of some manifold, finding a reduced integration measure is not trivial.

However there is one case where there is a natural inner product on the Hilbert space. This happens when the phase space is also Kähler and Ω is the Kähler form or some multiple thereof. In this case we can introduce local complex coordinates $(z^a, \bar{z}^{\bar{a}})$ and write

$$\Omega = \Omega_{a\bar{a}}\, \mathrm{d}z^a \wedge \mathrm{d}\bar{z}^{\bar{a}} \tag{4.18}$$

$a, \bar{a} = 1, 2...n$. The corresponding covariant derivatives are

$$\mathcal{D}_a = \partial_a - \mathrm{i}\mathcal{A}_a, \qquad \mathcal{D}_{\bar{a}} = \partial_{\bar{a}} - \mathrm{i}\mathcal{A}_{\bar{a}} \tag{4.19}$$

The characteristic of a Kähler manifold is the existence of a Kähler potential K such that

$$\mathcal{A}_a = -\frac{i}{2}\partial_a K, \qquad \mathcal{A}_{\bar{a}} = \frac{i}{2}\partial_{\bar{a}} K \qquad (4.20)$$

In this case, one can choose the holomorphic polarization (also referred to as the Bargmann polarization) defined by

$$\mathcal{D}_{\bar{a}}\Psi = (\partial_{\bar{a}} + \frac{1}{2}\partial_{\bar{a}} K)\Psi = 0 \qquad (4.21)$$

The solutions are the polarized wave functions ψ given by

$$\psi = \exp(-\tfrac{1}{2}K) \ \ F \qquad (4.22)$$

where F is a holomorphic function on M. The wave functions are thus holomorphic, apart from the prefactor involving the Kähler potential. In this case, ψ^* involves the antiholomorphic functions F^* and the product depends on all the phase space coordinates. Integration over all of phase space is acceptable and the inner product of the prequantum Hilbert space can be retained, may be up to a constant of proportionality, as the inner product of the true Hilbert space; specifically we have

$$\langle 1|2\rangle = \int d\sigma(M)\, e^{-K} \ \ F_1^* F_2 \qquad (4.23)$$

The cases where $M = T^*Q$ for some manifold Q and the Kähler case will cover most of the physical situations of interest to us.

4.4 Representation of Operators

Once the polarized wave functions are defined, the idea is to represent observables as linear operators on the wave functions as given by the prequantum differential operators. Let ξ be the Hamiltonian vector field corresponding to a function $f(q)$. If the commutator of ξ with any polarization vector field P_i is proportional to P_i itself, i.e., $[\xi, P_i] = C_i^j P_j$ for some functions C_i^j, then, evidently, ξ does not change the polarization; $\xi\Psi$ will obey the same polarization condition as Ψ. In this case the operator corresponding to $f(q)$ is given by $\mathcal{P}(f)$, but, of course, now acting on the wave functions in the chosen polarization.

The situation with operators which do not preserve the polarization is more complicated. There are many such operators of interest in any physical problem. For example, the Hamiltonian for a free nonrelativistic particle in one spatial dimension is $H = p^2/2m$, with the vector field $\xi_H = (p/m)\,(\partial/\partial x)$. If we choose the polarization which gives wave functions depending on x, namely, choose $P = (\partial/\partial p)$, then we find

$$[\xi_H, P] = -\frac{1}{m}\frac{\partial}{\partial x} \qquad (4.24)$$

We see that ξ_H does not preserve the polarization. The solution is also suggested by this example. We can define p^2 trivially by using the momentum-space wave functions, namely, ones corresponding to the polarization $(\partial/\partial x)$. It is possible to transform from one type of wave functions to the other; in this particular case, this is done by Fourier transformation. More generally, there are kernels, known as Blattner-Kostant-Sternberg (BKS) kernels, which map from one polarization to another [5]. Using this, we can define operators as follows. We carry out a canonical transformation on the wave functions by the vector field $t\,\xi_f$ where f is the function whose operator version we wish to find and t is a real parameter. The result is no longer in the same polarization, but we can transform back using an appropriate BKS kernel. The derivative of the result with respect to t at $t = 0$ will give the action of the operator. Equivalently, we can work out the form of the operator in a polarization which is preserved by the corresponding vector field and then transform to the required polarization using an appropriate BKS kernel.

4.5 Comments on the Measure of Integration, Corrected Operators, Etc

The problem of defining the measure of integration in a given polarization has implications, which necessitates a certain modified definition for operators. Fortunately, this will not be an issue for most of the examples we discuss later, but, nevertheless, a comment is in order at this stage. (For more detailed analysis, see [5, 7, 8].) To illustrate the problem, consider how we can show that $\mathcal{P}(f)$ is a symmetric operator, as in (4.6). The relevant integration-by-parts leads to a discrepancy $\partial_\mu \xi^\mu + \frac{1}{2}\xi^\mu \partial_\mu (\log \det \Omega)$ which is zero by virtue of the closure of Ω and ξ being a Hamiltonian vector field. However, the integration measure for the polarized wave functions is not given by Ω and hence this argument does not go through. Consider a real polarization and let the inner product be of the form

$$\langle 1|2\rangle = \int d^n x\, J\, \psi_1^* \,\psi_2 \qquad (4.25)$$

(We do not necessarily mean that x denotes coordinates of some configuration space, it is used as a generic notation here.) We then find

$$\int \psi_1^* \left(\mathcal{P}(f)\psi_2\right) - \int \left(\mathcal{P}(f)\psi_1\right)^* \psi_2 = i \int d^n x\, J\left[\partial \cdot \xi + \xi \cdot \partial \log J\right]\psi_1^*\,\psi_2$$
$$(4.26)$$

Clearly using $\mathcal{P}(f)$ to act on the polarized wave functions will not do. One strategy is to factorize J as $\bar{\sigma}\,\sigma$ where σ need not be real and consider $\psi\,\sigma$ in place of the wave function. The quantity σ behaves as the square root of the integration measure on the complement of the subspace defined by the polarization vector fields. For this reason, this way of considering $\psi\,\sigma$ directly, rather than ψ and then the measure of integration separately, is called the half-form quantization. We then modify the definition of the operator corresponding to f as[1]

$$\mathcal{P}(f)\,\psi\,\sigma = [(-i\xi\cdot\mathcal{D}+f)\,\psi]\,\sigma - \psi\,(iL_\xi\sigma)$$

$$-iL_\xi\sigma = -i\xi\cdot\partial\sigma - \frac{i}{2}\partial\cdot\xi\,\sigma \qquad (4.27)$$

With this definition, we can verify that

$$\int \mathrm{d}^n x\,\,\psi^*\bar{\sigma}\,\,[\mathcal{P}(f)\psi\,\sigma] = \int \mathrm{d}^n x\,\,[\mathcal{P}(f)\psi\,\sigma]^*\,\,\psi\,\sigma \qquad (4.28)$$

It is useful to consider the problem of the integration measure in some more detail. For this purpose, let us consider the Lagrangian submanifold defined by the polarization $\{P_i\}$. Let u^i denote the local coordinates on this submanifold. The coordinates q^μ on the submanifold can be considered as functions of u^i and obey equations of the form

$$(E^{-1})_i{}^k\,\frac{\partial q^\mu}{\partial u^k} = P_i{}^\mu \qquad (4.29)$$

The matrix of functions $E_i{}^k$ plays the role of frame fields for the subspace and we can define a volume measure of the form $(\det E)\,\mathrm{d}^n u$. In the inner product (4.25), the integrand $\psi_1^*\psi_2$ is independent of u^i. So, just as how one deals with the case of the functional integral for gauge theories, we can introduce a constraint $\delta^{(n)}(u)\,(\det E)^{-1}$ and integrate with the full Liouville measure. This will effectively remove the volume element $(\det E)\mathrm{d}^n u$ of the Lagrangian submanifold from the Liouville volume element. An alternative is to construct the antisymmetric tensor

$$\sigma_{-1}(P) = \frac{1}{n!}\epsilon^{i_1 i_2\cdots i_n}(E^{-1})_{i_1}^{k_1}(E^{-1})_{i_2}^{k_2}\cdots(E^{-1})_{i_n}^{k_n}\,\frac{\partial}{\partial u^{k_1}}\wedge\frac{\partial}{\partial u^{k_2}}\wedge\cdots\frac{\partial}{\partial u^{k_n}} \qquad (4.30)$$

The contraction of this with the Liouville volume form will remove the volume factor $(\det E)\,\mathrm{d}^n u$ of the Lagrangian submanifold defined by the P_i's. The expression for σ_{-1} given in (4.30) is in terms of local coordinates. We can extend this over the manifold M, so that σ_{-1} can be viewed as sections of an appropriate bundle $\delta_{-1}(P)$. Notice that if we make a transformation $P_i \to N_i{}^j\,P_j$ on the basis of polarization vectors, we have $(E^{-1})_i^k \to N_i^j(E^{-1})_j^k$ and

$$\sigma_{-1}(NP) = (\det N)\,\sigma_{-1}(P) \qquad (4.31)$$

---

[1] $L_\xi\sigma$ is again the Lie derivative of σ.

The reduced volume will have this transformation property. More generally, one can consider bundles $\delta_r(P)$ for which we have the transformation

$$\sigma_r(NP) = (\det N)^{-r}\, \sigma_r(P) \tag{4.32}$$

There are a couple of other properties obeyed by δ_r. For example, if we have dual spaces F and F^* (analogous to TM and T^*M), $\delta_r(F) = \delta_{-r}(F^*)$; further, from the transformation property (4.32), $\delta_r(F) \otimes \delta_s(F) = \delta_{r+s}(F)$. Also, in particular, $\delta_1(M)$ can be taken as the volume element for M.

In order to have a strategy which can work for all polarizations, including holomorphic (or partly holomorphic) ones, we will need to consider a "square root" of $\delta_{-1}(P)$, say $\delta_{-1/2}$. This bundle $\delta_{-1/2}$ is called a metaplectic structure on M. There are conditions on whether this square root can be defined consistently over the manifold, as we mention below. (We may think of $\delta_{1/2}$ as defining a volume on spinor frames.) The volume for the polarized subspace can be defined using $\delta_{-1/2}(P)$ and $\delta_{-1/2}(\bar{P})$ acting on $\delta_1(M)$. Let $W = (P \cup \bar{P})/(P \cap \bar{P})$. This can be shown to be a symplectic space with its own volume measure which will transform as σ_1. The general formula for the required integration measure is then

$$d\mu = \sigma_{-1/2}(P)\, \sigma_{-1/2}(\bar{P})\, \sigma_1(W)\, d\sigma\ (M) \tag{4.33}$$

For a real polarization, $P = \bar{P}$ and W is empty. Thus we get the result $\sigma_{-1}(P)\, d\sigma(M)$ which is the same as (4.25). The formula (4.33) factors out the effect of the directions defined by P. The two factors $\sigma_{-1/2}(P)$ and $\sigma_{-1/2}(\bar{P})$, which we denoted by σ and $\bar{\sigma}$ in Eqs. (4.27) and (4.28), are needed for the action of the operators as in (4.27).[2] In the integration measure, we can go back to the form $\delta^{(n)}(u)\det E^{-1}$ which is given by $\sigma_{-1}(P)$. If we consider holomorphic polarization, then $P \cup \bar{P} = M$ and $P \cap \bar{P} = \emptyset$, so that we get $\sigma_1(W) = \sigma_1(M)$, which gives another factor of $\sqrt{\det \Omega}$. This cancels with the $\sigma_{-1/2}$ factors retaining $d\sigma(M)$ as the volume. Thus we see that (4.33) will correctly reproduce the expected volume element.

An explicit formula for $\sigma_{-1/2}$ is not easy to construct, but this is not needed for most of the calculations. As seen from the statements above, the two factors of $\delta_{-1/2}$ often combine to produce an appropriate factor of δ_{-1}. However, the transformation rule is important in working out the consequences of using half-forms. As we see from (4.27), once we include such half-form factors, the definition of operators will have the corrections from $L_\xi \sigma_{-1/2}$. This can give a correction even when the operator ξ_f preserves the polarization. If ξ_f preserves the polarization, then we find

$$L_{\xi_f} P_i = [\xi_f, P_i] = C_i^j\, P_j \tag{4.34}$$

---

[2] By the way, $\sigma_r(\bar{P}) = \overline{\sigma_r(P)}$.

for some C_i^j. Since the polarization is preserved by ξ_f, the prequantum operator with the modification as in (4.27) can be used as the quantum version of f. Thus

$$P(f) = (-i\xi \cdot \mathcal{D} + f) - \frac{i}{2}\mathrm{Tr}\,C \qquad (4.35)$$

In comparison with (4.32), $N \approx 1 - iC$.

The condition for the existence of a metaplectic structure is essentially the same as the condition for the existence of spinors on the manifold, namely, the vanishing of the Stiefel-Whitney class; i.e., $\mathcal{H}^2(M, \mathbb{Z}_2) = 0$. (The metaplectic group is the covering group for the symplectic group, and $\delta_{-1/2}$ can be constructed using spinor frames.) If we have $\mathcal{H}^2(M, \mathbb{Z}_2) = 0$, then there can still be inequivalent $\delta_{-1/2}$ bundles, which are classified by $\mathcal{H}^1(M, \mathbb{Z}_2)$, exactly as for spinors.

The metaplectic structure gives a more formal and better way to address the issue of defining the integration measure for the inner product of the true wave functions and of having to modify the definition of operators corresponding to f as in (4.27). We will not go into this in any more detail here. The point is that, overall, while geometric quantization is very beautiful, it must be admitted that defining operators which do not preserve the polarization and defining an integration measure on the space of polarized wave functions are somewhat awkward and cumbersome. In what follows, we will be considering mostly the holomorphic polarization which avoids most of these issues.

Open Access This chapter is licensed under the terms of the Creative Commons Attribution 4.0 International License (http://creativecommons.org/licenses/by/4.0/), which permits use, sharing, adaptation, distribution and reproduction in any medium or format, as long as you give appropriate credit to the original author(s) and the source, provide a link to the Creative Commons license and indicate if changes were made.

The images or other third party material in this chapter are included in the chapter's Creative Commons license, unless indicated otherwise in a credit line to the material. If material is not included in the chapter's Creative Commons license and your intended use is not permitted by statutory regulation or exceeds the permitted use, you will need to obtain permission directly from the copyright holder.

Chapter 5
Topological Features of Quantization

Many of the topological properties of the phase space have an impact on the procedure of quantization and on key features of the quantum theory. We have already mentioned the Stiefel-Whitney class $\mathcal{H}^2(M, \mathbb{Z}_2)$ for the existence of the metaplectic structure and how $\mathcal{H}^1(M, \mathbb{Z}_2)$ will classify such structures. There are two other important features we will consider here. These are in relation to the first and second cohomology of the phase space. For further reading, see [5–7].

5.1 The Case of Nontrivial $\mathcal{H}^1(M, \mathbb{R})$

Consider first the case of $\mathcal{H}^1(M, \mathbb{R}) \neq 0$, which means that M admits one-forms, say A, which are closed but not exact. This implies that, for a given symplectic two-form Ω, we can have different symplectic potentials $\mathcal{A}$ and $\mathcal{A} + A$ which lead to the same Ω since A is closed, i.e., $dA = 0$. Now if A is exact, there is some globally defined function h on M such that $A = dh$. The function h is a canonical transformation and physical results will be unchanged. In fact an exact one-form is equivalent to $A = 0$ upon carrying out a canonical transformation. However, if A is closed but not exact, i.e., it is a nontrivial element of the cohomology $\mathcal{H}^1(M, \mathbb{R})$, then we cannot get rid of it by a canonical transformation. Locally we can still write $A = df$ for some f, but f will not be globally defined on M. Thus globally we cannot eliminate A.

Classical dynamics is defined by the equations of motion as in (3.2) which involves only Ω, not the symplectic potential $\mathcal{A}$. Thus this ambiguity in the choice of the symplectic potential due to nonzero $\mathcal{H}^1(M, \mathbb{R})$ will not affect the classical dynamics. In the quantum theory such A's do make a difference. This can be seen in terms of the action $\mathcal{S}$; for a path C, parametrized as $q^\mu(t)$ from a point a on M to a point b, the action is

$$\mathcal{S} = \int \mathrm{d}t \left(\mathcal{A}_\mu \frac{\mathrm{d}q^\mu}{\mathrm{d}t} - H \right) + \int_a^b A_\mu \mathrm{d}q^\mu \tag{5.1}$$

© The Author(s) 2024
V. P. Nair, *Geometric Quantization and Applications to Fields and Fluids*,
SpringerBriefs in Physics, https://doi.org/10.1007/978-3-031-65801-3_5

The action depends on the path but the contribution from A is topological. If we change the path slightly from C to C' with the end points fixed, we find, using Stokes' theorem,

$$\int_C A - \int_{C'} A = \oint_{C-C'} A = \int_\Sigma dA = 0 \tag{5.2}$$

where $C - C'$ is the path where we go from a to b along C and back from b to a along C'. (Since this is the return path, the orientation is reversed, hence the minus sign.) Σ is a surface in M with $C - C'$ as the boundary. The above result shows that the contribution from A is invariant under small changes of the path, which also explains why it does not contribute to the classical equations of motion, since the latter arise from extremization of the action under small variations. (The full action S does depend on the path.) In particular, the value of the integral of A is zero for closed paths so long as they are contractible; for then we can make a sequence of small deformations of the path (which do not change the value) and so the value of the integral will coincide with what it is for a path that is contracted to zero. In other words, the value will be zero.

If there are noncontractible loops, which is the case if $\mathcal{H}^1(M, \mathbb{R}) \neq 0$, then there can be nontrivial contributions arising from the integral of A around such loops. While this is irrelevant for the classical dynamics, in the quantum theory, it is e^{iS} which is important, so we need $e^{i \int A}$. (If one considers a path-integral formulation of the quantum theory, it is clear that e^{iS} is what is relevant. For the present discussion of an equal-time operator formulation, e^{iS} is again the relevant quantity as it determines the phases of the wave functions, which can be measurable via interference. Hence a nontrivial $e^{i \int A}$ has physical consequences.)

Assume for simplicity that $\mathcal{H}^1(M, \mathbb{R})$ has only one nontrivial element (say α) up to addition of trivial terms and multiplicative factors. Then there is only one topologically distinct noncontractible loop apart from multiple traversals of the same. Let $A = \theta \alpha$ where θ is a constant and α is normalized to unity along the noncontractible loop for going round once. For all paths which include n traversals of the loop, we find

$$\exp\left(i \oint A\right) = \exp\left(i\theta \oint \alpha\right) = \exp(i\theta n) \tag{5.3}$$

Notice that a shift $\theta \to \theta + 2\pi$ does not change this value, so that we may restrict θ to be in the interval zero to 2π. Putting this back into the action (5.2), we see that, as a function over all paths, the action has an extra parameter θ. Thus the ambiguity in the choice of the symplectic potential due to $\mathcal{H}^1(M, \mathbb{R}) \neq 0$ leads to an extra parameter θ which is needed to fully characterize the quantum theory. Since θ is in the interval 0 to 2π, we may regard $A = \theta \alpha$ as an element of $\mathcal{H}^1(M, \mathbb{R})/\mathcal{H}^1(M, \mathbb{Z})$. If $\mathcal{H}^1(M, \mathbb{R})$ has more than one distinct element, there are more distinct paths possible and there can be more parameters like θ. Such parameters are generally called vacuum angles.

It is now easy to see these results in terms of wave functions. The relevant covariant derivatives are of the form $\mathcal{D}_\mu \Psi = (\partial_\mu - i\mathcal{A}_\mu - iA_\mu)\Psi$. We can write

$$\Psi(q) = \exp\left(i \int_a^q A\right) \Phi(q) \tag{5.4}$$

where the lower limit of the integral is some fixed point a. By using this in the covariant derivative, we see that A is removed from $\mathcal{D}_\mu$ in terms of action on Φ; $\mathcal{D}_\mu$ acting on Φ is then just $\mathcal{D}_\mu = \partial_\mu - i\mathcal{A}_\mu$. The redefinition of the wave functions in (5.4) is like a canonical transformation, except that the relevant factor $\exp\left(i \int_0^q A\right)$ is not single valued. As we go around a closed noncontractible curve, it can give a phase $e^{i\theta}$. Since Ψ is single-valued, this means that Φ must have a compensating phase factor; Φ is not single-valued but must give a specific phase labelled by θ. Thus we can get rid of A from the covariant derivatives, and hence from various operator formulae, by taking the wave functions to be the Φ's related to the Ψ's as in (5.4). But diagonalizing the Hamiltonian on such Φ's can give results which depend on the angle θ, since the Φ's must pick up a phase $e^{-i\theta}$ for each traversal of the noncontractible loop.

The θ-vacua in a nonabelian gauge theory is an example of this kind of topological feature. The description of particles of fractional statistics in two spatial dimensions is another example.

5.2　The Case of Nontrivial $\mathcal{H}^2(M, \mathbb{R})$

We now turn to the second topological feature mentioned earlier, namely the case of $\mathcal{H}^2(M, \mathbb{R}) \neq 0$. This means that there are closed two-forms on M which are not exact. Correspondingly, there are closed two-surfaces which are not the boundaries of any three-dimensional region, i.e., there exists noncontractible closed two-surfaces. In general, elements of $\mathcal{H}^2(M, \mathbb{R})$ integrated over such noncontractible two-surfaces will not be zero. If the symplectic two-form Ω is some nontrivial element, or it has a part which is a nontrivial element, of $\mathcal{H}^2(M, \mathbb{R})$, then the symplectic potential $\mathcal{A}$ cannot be globally defined. This is easily seen from the following argument. Consider the integral of Ω over a noncontractible two-surface Σ,

$$I(\Sigma) = \int_\Sigma \Omega \tag{5.5}$$

First of all, this is a topological invariant, for if Σ' is a small deformation of Σ, then

$$I(\Sigma) - I(\Sigma') = \int_{\Sigma - \Sigma'} \Omega = \int_V d\Omega = 0 \tag{5.6}$$

where V is a three-dimensional volume with the two surfaces $\Sigma - \Sigma'$ as the boundary. This shows that the integral of Ω is invariant under small deformations of the surface over which it is integrated. If we could write Ω as $d\mathcal{A}$ for some $\mathcal{A}$ which is globally defined on Σ then clearly $I(\Sigma)$ is zero by Stokes' theorem. Thus if $I(\Sigma)$ is nonzero, we must conclude that there is no potential $\mathcal{A}$ which is globally defined on Σ. We have to use different choices for $\mathcal{A}$ in different coordinate patches on M and have transition functions relating the $\mathcal{A}$'s in the overlap regions. But we must have the same Ω on a given overlap region whether we use the $\mathcal{A}$ for one patch or the $\mathcal{A}$ for the other patch to calculate it. Thus the transition functions on overlap regions must be canonical transformations (or gauge transformations on $\mathcal{A}$).

As an example, consider a closed noncontractible two-sphere, or any smooth deformation of it, which may be a subspace of M. We can cover it with two coordinate patches corresponding to the two hemispheres, denoted N and S as usual. The symplectic potential is represented by $\mathcal{A}_N$ and $\mathcal{A}_S$ respectively. On the equatorial overlap region, they are connected by

$$\mathcal{A}_N = \mathcal{A}_S + d\Lambda \tag{5.7}$$

where Λ is a function defined on the overlap region. It gives the canonical transformation between the two $\mathcal{A}$'s.

The symplectic potential $\mathcal{A}$ is what is needed in setting up the quantum theory. And since canonical transformations are represented as unitary transformations on the wave functions, we see that we must also have a Ψ_N for the patch N and a Ψ_S for the patch S. On the equator they must be related by the canonical transformation, which from (4.1), is given as

$$\Psi_N = \exp(i\,\Lambda)\ \Psi_S \tag{5.8}$$

Now consider the integral of $d\Lambda$ over the equator E, which is a closed curve being the boundary of either N or S. From (5.7) this is given as

$$\Delta\Lambda = \oint_E d\Lambda = \int_E \mathcal{A}_N - \int_E \mathcal{A}_S = \int_{\partial N} \mathcal{A}_N + \int_{\partial S} \mathcal{A}_S$$
$$= \int_N \Omega + \int_S \Omega = \int_\Sigma \Omega \tag{5.9}$$

(In the second step, we reverse the sign for the S-term because E considered as the boundary of S has the opposite orientation compared to it being the boundary of N.) The above equation shows that the change of Λ as we go around the equator once, namely $\Delta\Lambda$, is nonzero if $I(\Sigma)$ is nonzero. In other words, Λ is not single-valued on the equator. But the wave function must be single-valued. From (5.8), we see that this can be achieved if $\exp(i\Delta\Lambda) = 1$ or if $\Delta\Lambda = 2\pi n$ for some integer n. Combining with (5.9), this can be stated as a topological quantization rule implied by the single-valuedness of wave functions in the quantum theory,

$$\int_\Sigma \Omega = 2\pi n \tag{5.10}$$

The integral of the symplectic two-form Ω on closed noncontractible two-surfaces must be quantized as 2π times an integer. (Or we may say that Ω must belong to an integral cohomology class of M.) Notice that $\mathcal{A}$ only sees $d\Lambda$ as we go from one patch to the other, and the transition condition on the wave functions only involve $\exp(i \Lambda)$ as in (5.8). Both $d\Lambda$ and $\exp(i \Lambda)$ are single-valued if the condition (5.10) is satisfied, so there is no difficulty for any observable quantity in the quantum theory.

We have given the argument for surfaces which are deformations of a two-sphere, but a similar argument can be made for general noncontractible two-surfaces. The quantization condition (5.10) on Ω is quite general: The integral of Ω over any closed two-cycle in M must be 2π times an integer.

The quintessential example of this kind of topological feature is the motion of a charged particle in the field of a magnetic monopole. The condition (5.10) is then the famous Dirac quantization condition. The Wess-Zumino terms occuring in many field theories are another example.

5.3 Summary of Holomorphic Polarization and Quantization

Since we will be using geometric quantization with holomorphic polarization in some of the examples later, this is a good point to summarize the key features of quantization using the holomorphic polarization.

1. We need a phase space which is also Kähler; the symplectic two-form must be a multiple of the Kähler form.
2. The prequantum wave functions are sections of a bundle which is the product of the holomorphic line bundle with curvature equal to the symplectic form and a half-form bundle. (The existence of the half-form bundle requires the vanishing of the Stiefel-Whitney class as mentioned earlier.)
3. The true wave functions are obtained by imposing the polarization condition, which, for the holomorphic polarization is $\mathcal{D}_{\bar{a}} \Psi = 0$.
4. The inner product of the prequantum Hilbert space, which is essentially square integrability on the phase space with the Liouville measure of integration, is retained as the inner product on the true Hilbert space in the holomorphic polarization.
5. The operator corresponding to an observable $f(q)$ which preserves the chosen polarization is given by the prequantum operator $\mathcal{P}(f)$ acting on the true (polarized) wave functions. The half-form part of the wave functions, while not important for the integration measure in the holomorphic polarization, can modify the operators as in (4.27) or (4.35).
6. For observables which do not preserve the polarization, one has to construct infinitesimal unitary transformations whose classical limits are the required canonical transformations.

7. If $\mathcal{H}^1(M, \mathbb{R})$ is not zero, then there are inequivalent $\mathcal{A}$'s for the same Ω and we need extra angular parameters to specify the quantum theory completely.
8. If the phase space M has noncontractible two-surfaces, then the integral of Ω over any of these surfaces must be quantized in units of 2π.

Problem

5.1 For a particle moving on a circle with coordinate θ, $d\theta/(2\pi)$ is an element of $\mathcal{H}^1(M)$. Consider the action

$$\mathcal{S} = \int dt \left[\tfrac{1}{2}\dot\theta^2 + \frac{\alpha}{2\pi}\dot\theta \right]$$

Obtain the energy eigenvalues to show how they depend on the vacuum angle α.

Open Access This chapter is licensed under the terms of the Creative Commons Attribution 4.0 International License (http://creativecommons.org/licenses/by/4.0/), which permits use, sharing, adaptation, distribution and reproduction in any medium or format, as long as you give appropriate credit to the original author(s) and the source, provide a link to the Creative Commons license and indicate if changes were made.

The images or other third party material in this chapter are included in the chapter's Creative Commons license, unless indicated otherwise in a credit line to the material. If material is not included in the chapter's Creative Commons license and your intended use is not permitted by statutory regulation or exceeds the permitted use, you will need to obtain permission directly from the copyright holder.

Chapter 6
Coherent States, the Two-Sphere and G/H Spaces

We will now consider some examples of geometric quantization. Specifically, we discuss coherent states in flat space, on the two-sphere and on the complex projective space, using local coordinates, homogencous coordinates and a group-theoretic formulation [5–7, 9]. Quantization of Kähler spaces of the G/H-type and the use of index theorems to calculate the dimension of the Hilbert space will also be briefly outlined.

6.1 Coherent States

We will start with the simplest case of coherent states for a one-dimensional quantum system to illustrate how the ideas of geometric quantization take concrete form. In one spatial dimension, $\Omega = \mathrm{d}p \wedge \mathrm{d}x = \mathrm{i}(\mathrm{d}z \wedge \mathrm{d}\bar{z})/\kappa$, where $z, \bar{z} = \kappa\,(p \pm \mathrm{i}x)/\sqrt{2}$. (We introduce a parameter κ which will be useful for later considerations.) Choose

$$\mathcal{A} = \frac{\mathrm{i}}{2\kappa}(z\,\mathrm{d}\bar{z} - \bar{z}\,\mathrm{d}z) \tag{6.1}$$

The space has the Kähler property, with the Kähler potential $K = \bar{z}z/\kappa$. The covariant derivatives corresponding to (6.1) are

$$\mathcal{D}_z = \partial_z - \frac{\bar{z}}{2\kappa}, \qquad \mathcal{D}_{\bar{z}} = \partial_{\bar{z}} + \frac{z}{2\kappa} \tag{6.2}$$

Holomorphic polarization corresponds to $P = \partial/\partial\bar{z}$, so that the polarization condition on the prequantum wave functions is

$$\mathcal{D}_{\bar{z}}\Psi = \left(\partial_{\bar{z}} + \frac{z}{2\kappa}\right)\Psi = 0 \tag{6.3}$$

© The Author(s) 2024
V. P. Nair, *Geometric Quantization and Applications to Fields and Fluids*,
SpringerBriefs in Physics, https://doi.org/10.1007/978-3-031-65801-3_6

The solutions of this equation are of the form

$$\Psi = e^{-\frac{1}{2}(z\bar{z}/\kappa)} \varphi(z) \tag{6.4}$$

where $\varphi(z)$ is holomorphic in z. The Hamiltonian vector fields corresponding to $z, \bar{z}$ are

$$z \longleftrightarrow -i\kappa\frac{\partial}{\partial\bar{z}}, \qquad \bar{z} \longleftrightarrow i\kappa\frac{\partial}{\partial z} \tag{6.5}$$

These commute with $P = \partial/\partial\bar{z}$ and so are polarization-preserving. The prequantum operators corresponding to these are

$$\mathcal{P}(z) = -i(-i\kappa)\left(\frac{\partial}{\partial\bar{z}} + \frac{z}{2\kappa}\right) + z = -\kappa\frac{\partial}{\partial\bar{z}} + \tfrac{1}{2}z$$

$$\mathcal{P}(\bar{z}) = -i(i\kappa)\left(\frac{\partial}{\partial z} - \frac{\bar{z}}{2\kappa}\right) + \bar{z} = \kappa\frac{\partial}{\partial z} + \tfrac{1}{2}\bar{z} \tag{6.6}$$

In terms of their action on the functions $\varphi(z)$ in (6.4), corresponding to Ψ's obeying the polarization condition, we define the operator versions of z and $\bar{z}$ by

$$\mathcal{P}(z)\,\Psi = e^{-\frac{1}{2}(\bar{z}z/\kappa)}\,\mathcal{O}(z)\varphi(z), \qquad \mathcal{P}(\bar{z})\,\Psi = e^{-\frac{1}{2}(\bar{z}z/\kappa)}\,\mathcal{O}(\bar{z})\varphi(z) \tag{6.7}$$

so that

$$\mathcal{O}(z)\,\varphi(z) = z\,\varphi(z)$$

$$\mathcal{O}(\bar{z})\,\varphi(z) = \kappa\frac{\partial\varphi}{\partial z} \tag{6.8}$$

The inner product for the $\varphi(z)$'s is

$$\langle 1|2 \rangle = \int i\frac{dz \wedge d\bar{z}}{2\pi\kappa}\, e^{-z\bar{z}/\kappa}\, \varphi_1^*\, \varphi_2 \tag{6.9}$$

A basis for the Hilbert space of states is given by

$$\psi_n = e^{-\frac{1}{2}(z\bar{z}/\kappa)}\frac{z^n}{\kappa^{\frac{n}{2}}\sqrt{n!}} \equiv \langle z|n \rangle \tag{6.10}$$

What we have obtained is the standard coherent state (or Bargmann) realization of the Heisenberg algebra.

It is illuminating to consider the quantization of the function $\bar{z}z$. The vector field corresponding to this is $\xi = i\kappa(z\partial_z - \bar{z}\partial_{\bar{z}})$. The prequantum operator for this is easily seen to be $z\partial_z$ acting on $\varphi(z)$. For the polarization we have chosen,

$$[\xi, \partial_{\bar{z}}] = i\kappa \, \partial_{\bar{z}} \tag{6.11}$$

Thus ξ preserves polarization and we can identify $C = i\kappa$ in comparing with (4.34). The operator corresponding to $\bar{z}z$, including the metaplectic correction, is thus

$$\mathcal{O}(\bar{z}z) = \kappa \left(z\frac{\partial}{\partial z} + \frac{1}{2} \right) \tag{6.12}$$

For most of what follows we will set $\kappa = 1$.

6.2 Quantizing the Two-Sphere

We now consider the example of the phase space being a two-sphere S^2. This space can be considered as $\mathbb{CP}^1$, the complex projective space in one (complex) dimension. It is a Kähler manifold. We may also regard S^2 as $SU(2)/U(1)$, a point of view which is useful for generalization later. We will consider quantization of the two-sphere first in local coordinates, then using homogeneous coordinates for $\mathbb{CP}^1$, and then from the group theory point of view.

6.2.1 Quantization Using Local Coordinates

We introduce local complex coordinates for $\mathbb{CP}^1$ as $z = x + iy, \bar{z} = x - iy$, the standard Kähler two-form is given by

$$\omega = i \frac{dz \wedge d\bar{z}}{(1 + z\bar{z})^2} \tag{6.13}$$

These coordinates can be related to an embedding of S^2 in $\mathbb{R}^3$ via

$$X_1 = \frac{z + \bar{z}}{(1 + z\bar{z})}, \qquad X_2 = \frac{i(z - \bar{z})}{(1 + z\bar{z})}, \qquad X_3 = \frac{1 - z\bar{z}}{(1 + z\bar{z})} \tag{6.14}$$

so that we may view $z, \bar{z}$ as the coordinates of a plane onto which the sphere is stereographically projected. The metric is given by $ds^2 = e^1 e^1 + e^2 e^2$ where the frame fields are

$$e^1 = \frac{dx}{1 + z\bar{z}}, \qquad\qquad e^2 = \frac{dy}{1 + z\bar{z}} \qquad (6.15)$$

The Riemannian curvature is $R^1_2 = 4\,e^1 \wedge e^2$, giving the Euler number

$$\chi = \int \frac{R_{12}}{2\pi} = 2 \qquad (6.16)$$

The phase space has nonzero $\mathcal{H}^2(M, \mathbb{R})$ with its generating element given by the Kähler two-form, which is also proportional to the volume form for S^2. As the discussion which led to (5.10) showed, the symplectic two-form must belong to an integral cohomology class of M to be able to quantize properly. So we consider the symplectic form

$$\Omega = n\,\omega = i\,n\,\frac{dz \wedge d\bar{z}}{(1 + z\bar{z})^2} = i\,\partial\,\bar{\partial}\,K, \qquad K = n\log(1 + z\bar{z}) \qquad (6.17)$$

where n is an integer. In this case, we can verify by direct evaluation of the integral that

$$\int_M \Omega = 2\pi n \qquad (6.18)$$

as required by the quantization condition. In (6.17), K is the Kähler potential for Ω. Classically the Poisson bracket of two functions F and G on the phase space is given by

$$\begin{aligned}
\{F, G\} &= \Omega^{\mu\nu}\,\partial_\mu F\,\partial_\nu G \\
&= \frac{i}{n}(1 + z\bar{z})^2 \left(\frac{\partial F}{\partial z} \frac{\partial G}{\partial \bar{z}} - \frac{\partial F}{\partial \bar{z}} \frac{\partial G}{\partial z} \right)
\end{aligned} \qquad (6.19)$$

Turning to the quantization, first of all, the symplectic potential corresponding to the Ω in (6.17) can be taken as

$$\mathcal{A} = i\frac{n}{2} \left[\frac{z\,d\bar{z} - \bar{z}\,dz}{(1 + z\bar{z})} \right] \qquad (6.20)$$

The covariant derivatives, which are given by $\partial - i\mathcal{A}$, are

$$\begin{aligned}
\mathcal{D}_z &= \partial_z - i\mathcal{A}_z = \partial_z - \frac{n}{2}\frac{\bar{z}}{1 + z\bar{z}} \\
\mathcal{D}_{\bar{z}} &= \partial_{\bar{z}} - i\mathcal{A}_{\bar{z}} = \partial_{\bar{z}} + \frac{n}{2}\frac{z}{1 + z\bar{z}}
\end{aligned} \qquad (6.21)$$

The holomorphic polarization condition is

$$
\mathcal{D}_{\bar{z}}\Psi = (\partial_{\bar{z}} - i\mathcal{A}_{\bar{z}})\Psi = \left[\partial_{\bar{z}} + \frac{n}{2}\frac{z}{(1+z\bar{z})}\right]\Psi = 0 \tag{6.22}
$$

This can be solved as

$$
\Psi = \exp\left(-\frac{n}{2}\log(1+z\bar{z})\right) f(z) \tag{6.23}
$$

where $f(z)$ is a holomorphic function of z. Notice that we have a factor $\exp(-\frac{1}{2}K)$ as expected. The inner product is given by

$$
\langle 1|2\rangle = i c \int \frac{dz \wedge d\bar{z}}{2\pi(1+z\bar{z})^{n+2}} f_1^* f_2 \tag{6.24}
$$

Here c is an overall constant, which can be absorbed into the normalization factors for the wave functions. Since $f(z)$ in (6.23) is holomorphic, we can see that a basis of nonsingular wave functions is given by $f(z) = 1, z, z^2, \ldots, z^n$; higher powers of z will not have finite norm. The dimension of the Hilbert space is thus $(n+1)$. We could have seen that this dimension would be finite from the semiclassical estimate of the number of states as the phase volume. Since the phase volume is finite for $M = S^2$, the dimension of the Hilbert space should be finite.

An orthonormal basis for the wave functions may be taken to be

$$
f_k(z) = \left[\frac{n!}{k!\,(n-k)!}\right]^{\frac{1}{2}} z^k \tag{6.25}
$$

with the inner product

$$
\langle 1|2\rangle = i(n+1) \int \frac{dz \wedge d\bar{z}}{2\pi(1+z\bar{z})^{n+2}} f_1^* f_2 \tag{6.26}
$$

Here we have chosen the parameter c in (6.24) such that the trace of the identity operator is the dimension of the Hilbert space, equal to $n+1$.

Consider now the vector fields

$$
\xi_+ = i\left(\frac{\partial}{\partial\bar{z}} + z^2\frac{\partial}{\partial z}\right), \quad \xi_- = -i\left(\frac{\partial}{\partial z} + \bar{z}^2\frac{\partial}{\partial\bar{z}}\right), \quad \xi_3 = i\left(z\frac{\partial}{\partial z} - \bar{z}\frac{\partial}{\partial\bar{z}}\right) \tag{6.27}
$$

It is easily verified that these are the standard $SU(2)$ isometries of the sphere. The Lie commutator of the ξ's give the $SU(2)$ algebra. Further, these are Hamiltonian vector fields corresponding to the functions

$$
J_+ = -n\frac{z}{1+z\bar{z}}, \quad J_- = -n\frac{\bar{z}}{1+z\bar{z}}, \quad J_3 = -\frac{n}{2}\left(\frac{1-z\bar{z}}{1+z\bar{z}}\right) \tag{6.28}
$$

The prequantum operators $-i\xi \cdot \mathcal{D} + J$ corresponding to these functions are

$$\mathcal{P}(J_+) = \left(z^2\partial_z - \frac{nz}{2}\frac{2+\bar{z}z}{1+\bar{z}z}\right) - i\xi_+^{\bar{z}}\mathcal{D}_{\bar{z}}$$

$$\mathcal{P}(J_-) = \left(-\partial_z - \frac{n}{2}\frac{\bar{z}}{1+z\bar{z}}\right) - i\xi_-^{\bar{z}}\mathcal{D}_{\bar{z}}$$

$$\mathcal{P}(J_3) = \left(z\partial_z - \frac{n}{2}\frac{1}{1+z\bar{z}}\right) - i\xi_3^{\bar{z}}\mathcal{D}_{\bar{z}} \tag{6.29}$$

Acting on the polarized wave functions, $\mathcal{D}_{\bar{z}}$ in these expressions will give zero. Writing Ψ as in (6.23), we can then work out the action of the operators on the holomorphic wave functions $f(z)$, by moving the derivatives through the $e^{-\frac{1}{2}K}$ factor. We then find

$$\hat{J}_+ f = (z^2\partial_z - nz)f$$

$$\hat{J}_- f = (-\partial_z)f$$

$$\hat{J}_3 f = (z\partial_z - \tfrac{1}{2}n)f \tag{6.30}$$

If we define $j = n/2$, which is therefore half-integral, we see that the operators given above correspond to a unitary irreducible representation of $SU(2)$ with $J^2 = j(j+1)$ and dimension $n+1 = 2j+1$. Notice that there is only one representation here and it is fixed by the choice of the symplectic form Ω. In other words, the quantization of the two-sphere with the symplectic form (6.17) gives one unitary irreducible representation of $SU(2)$ with $j = n/2$.

6.2.2   Quantization Using Homogeneous Coordinates

The complex coordinates we used are only local coordinates valid in a coordinate patch around $z = 0$; strictly speaking we need at least another coordinate patch with a different choice of coordinates to describe the sphere in a nonsingular way. This second patch would be needed around $z = \infty$, corresponding to the south pole of the sphere S^2 in the stereographic projection (6.14). It did not matter too much in what we did so far, because the potential coordinate singularity is basically a point with zero measure.

A more global approach is to use the homogeneous coordinates of the sphere viewed as $\mathbb{CP}^1$. Recall that the complex projective space $\mathbb{CP}^k$ is defined by $(k+1)$ complex coordinates $(u_1, u_2, \ldots, u_{k+1}) \in \mathbb{C}^{k+1}$ with the identification $(u_1, u_2, \ldots, u_{k+1}) \sim \lambda(u_1, u_2, \ldots, u_{k+1})$, for any complex nonzero λ, $\lambda \in \mathbb{C} - \{0\}$. Thus, for $\mathbb{CP}^1$, we will need two u's which we may think of as a two-component spinor u_α, $\alpha = 1, 2$, with the identification $u_\alpha \sim \lambda u_\alpha$. We also define $\bar{u}_1 = u_2^*$, $\bar{u}_2 = -u_1^*$

or $\bar{u}_\alpha = \epsilon_{\alpha\beta} u_\beta^*$, where $\epsilon_{\alpha\beta} = -\epsilon_{\beta\alpha}$, $\epsilon_{12} = 1$. The metric corresponding to the frames (6.15) is

$$ds^2 = \left[\frac{d\bar{u} \cdot du}{\bar{u} \cdot u} - \frac{\bar{u} \cdot du \, d\bar{u} \cdot u}{(\bar{u} \cdot u)^2} \right] \tag{6.31}$$

This is known as the Fubini-Study metric. The symplectic form corresponding to (6.17) is

$$\Omega = -in \left[\frac{du \cdot d\bar{u}}{\bar{u} \cdot u} - \frac{\bar{u} \cdot du \, u \cdot d\bar{u}}{(\bar{u} \cdot u)^2} \right] \tag{6.32}$$

where the notation is $u \cdot v = u_\alpha v_\beta \epsilon_{\alpha\beta}$. This means that $\bar{u} \cdot v = u^\dagger v = u_1^* v_1 + u_2^* v_2$.[1] It is easily checked that $\Omega(\lambda u) = \Omega(u)$; it is invariant under $u \to \lambda u$ and hence is properly defined on $\mathbb{CP}^1$ rather than on $\mathbb{C}^2$. The choice of $u_2/u_1 = z$ leads to the previous local parametrization; this is valid around $u_1 \neq 0$. We can use another coordinate patch with the local coordinates $w = u_1/u_2$. These two patches will correspond to the north and south hemispheres of the sphere, in the stereographic projection.

The symplectic potential corresponding to (6.32) is

$$\mathcal{A} = -i\frac{n}{2} \left[\frac{u \cdot d\bar{u} - du \cdot \bar{u}}{\bar{u} \cdot u} \right] \tag{6.33}$$

Directly from the above expression we see that

$$\mathcal{A}(\lambda u) = \mathcal{A}(u) + d \left(i\frac{n}{2} \log(\bar{\lambda}/\lambda) \right) \tag{6.34}$$

This means that $\mathcal{A}$ cannot be written as a globally defined form on $\mathbb{CP}^1$ since it is not invariant under the needed identification $u_\alpha \sim \lambda u_\alpha$. This is to be expected because $\int \Omega \neq 0$ and hence we cannot have a globally defined potential on $\mathbb{CP}^1$. From the transformation law (6.34) and (4.1), we see that the prequantum wave functions must transform as

$$\Psi(\lambda u, \bar{\lambda}\bar{u}) = \Psi(u, \bar{u}) \, \exp\left[\frac{n}{2} \log(\lambda/\bar{\lambda}) \right] \tag{6.35}$$

The polarization condition for the wave functions becomes

$$\left[\frac{\partial}{\partial \bar{u}_\alpha} - \frac{n}{2} \frac{u_\beta \, \epsilon_{\beta\alpha}}{\bar{u} \cdot u} \right] \Psi = 0 \tag{6.36}$$

The solution to this condition is

$$\Psi = \exp\left(-\frac{n}{2} \log(\bar{u} \cdot u) \right) f(u) \tag{6.37}$$

---

[1] In (6.32), a wedge product is implied, while there is no wedge product in (6.31).

Combining this with (6.35), we see that the holomorphic functions $f(u)$ should behave as

$$f(\lambda u) = \lambda^n f(u) \tag{6.38}$$

Therefore $f(u)$ must have n u's and hence is of the form

$$f(u) = \sum_{\alpha's} C^{\alpha_1 \cdots \alpha_n} u_{\alpha_1} \cdots u_{\alpha_n} \tag{6.39}$$

Because of the symmetry of the indices, there are $n+1$ independent functions, as before. There is a natural linear action of $SU(2)$ on the u, $\bar{u}$ given by

$$u'_\alpha = U_{\alpha\beta} u_\beta, \qquad\qquad \bar{u}'_\alpha = U_{\alpha\beta} \bar{u}_\beta \tag{6.40}$$

where $U_{\alpha\beta}$ form a (2×2) $SU(2)$ matrix. The corresponding generators are the J_a we have constructed in (6.29) and (6.30). We have thus recovered all the previous results in a more global way.

6.2.3 Group Theoretic Version

Equation (3.5) relating the action and the symplectic potential $\mathcal{A}$ shows that the potential of interest to us, namely, (6.20) can be obtained from the action

$$\mathcal{S} = \mathrm{i}\frac{n}{2} \int \mathrm{d}t \, \frac{z\dot{\bar{z}} - \bar{z}\dot{z}}{1 + z\bar{z}} \tag{6.41}$$

where the overdot denotes differentiation with respect to time. This action may be written as [10, 11]

$$\mathcal{S} = \mathrm{i}\frac{n}{2} \int \mathrm{d}t \, \mathrm{Tr}(\sigma_3 \, g^{-1}\dot{g}) \tag{6.42}$$

where g is an element of $SU(2)$ written as a (2×2)-matrix, $g = \exp(\mathrm{i}\,(\sigma_i/2)\theta_i)$ and $\sigma_i, i = 1, 2, 3$, are the Pauli matrices, given explicitly as

$$\sigma_1 = \begin{pmatrix} 0 & 1 \\ 1 & 0 \end{pmatrix}, \qquad \sigma_2 = \begin{pmatrix} 0 & -\mathrm{i} \\ \mathrm{i} & 0 \end{pmatrix}, \qquad \sigma_3 = \begin{pmatrix} 1 & 0 \\ 0 & -1 \end{pmatrix} \tag{6.43}$$

In the action (6.42), the dynamical variable is thus an element of $SU(2)$. There are many ways to parametrize the group element, corresponding to local coordinates on the group viewed as a Riemannian manifold. One convenient parametrization is given by

$$g = \frac{1}{\sqrt{1 + z\bar{z}}} \begin{pmatrix} 1 & z \\ -\bar{z} & 1 \end{pmatrix} \begin{bmatrix} e^{i\frac{\theta}{2}} & 0 \\ 0 & e^{-i\frac{\theta}{2}} \end{bmatrix} \tag{6.44}$$

If this is used in (6.42), we get (6.41).

In the action (6.42), if we make a transformation $g \to gh, h = \exp(i\sigma_3\varphi)$, we get

$$S \to S - n \int dt \, \dot{\varphi} \tag{6.45}$$

The extra term is a boundary term and does not affect the equations of motion. (It is for this same reason that θ in (6.44) does not appear in (6.41).) Since equations of motion do not depend on θ, we see that classically the dynamics is actually restricted to $SU(2)/U(1) = S^2$.

Even though classical dynamics is restricted to $SU(2)/U(1)$, the boundary term in (6.45) does have an effect in the quantum theory. Consider choosing $\varphi(t)$ such that $\varphi(-\infty) = 0$ and $\varphi(\infty) = 2\pi$. In this case $h(-\infty) = h(\infty) = 1$ giving a closed loop in the $U(1)$ subgroup of $SU(2)$ defined by the σ_3-direction. For this choice of $h(t)$, the action changes by $-2\pi n$. However, e^{iS} remains single-valued since n is chosen to be an integer, and, even in the quantum theory, the extra $U(1)$ degree of freedom is consistently removed. If the coefficient were not an integer, this would not be the case and we would have inconsistencies in the quantum theory. The quantization of the coefficient to an integral value, already seen in (6.18), is seen again from a slightly different point of view.

We can now move ahead and complete the quantization. The canonical one-form is obtained from S as

$$\mathcal{A} = i\frac{n}{2}\text{Tr}(\sigma_3 \, g^{-1}dg) \tag{6.46}$$

The corresponding two-form is given by

$$\Omega = -i\frac{n}{2}\text{Tr}(\sigma_3 \, g^{-1}dg \, g^{-1}dg) \tag{6.47}$$

The prequantum wave functions are sections of a bundle on $SU(2)/U(1)$. Let us start with functions on $SU(2)$. A function on $SU(2)$ may be written as a linear combination of the representation matrices $\mathcal{D}^{(j)}_{ab}(g)$ as

$$\Psi = \sum_j \sum_{a,b} C^{(j)}_{ab} \mathcal{D}^{(j)}_{ab}(g) = \sum_j \sum_{a,b} C^{(j)}_{ab} \langle a|e^{i\hat{J}_i\theta_i}|b\rangle \tag{6.48}$$

where $\hat{J}_i$ is the angular momentum or $SU(2)$ generator in an arbitrary representation. (The matrices $\mathcal{D}^{(j)}_{ab}(g)$ are also known as the Wigner $\mathcal{D}$-functions.) Consider the transformation $g \to gh, h = \exp(i\frac{\sigma_3}{2}\theta)$; the change in $\mathcal{A}$ is given by $\mathcal{A} \to \mathcal{A} - (n/2)\,d\theta$. Since $\sigma_3/2$ corresponds to $\hat{J}_3$ in an arbitrary representation, this implies that the wave functions must obey

$$\Psi\left(g\,e^{\mathrm{i}\hat{J}_3\theta}\right) = \Psi(g)\,\exp\left(-\frac{\mathrm{i}\,n}{2}\theta\right) \tag{6.49}$$

This identifies the J_3-eigenvalue of the state $|b\rangle$ in (6.48) as $-n/2$, so that we can write $|b\rangle = |j, -\frac{n}{2}\rangle$.

We have considered translations of g on the right by $h \in U(1)$. The remaining generators for the right action are $R_\pm = R_1 \pm \mathrm{i} R_2$, where R_i is defined by

$$R_i\, g = g\,\frac{\sigma_i}{2} \tag{6.50}$$

The combinations $R_\pm$ are complex and conjugate to each other. We can take R_- as the polarization condition, requiring the wave functions to obey

$$R_-\,\Psi = R_- \sum_j \sum_{a,b} C_{ab}^{(j)}\,\langle a|e^{\mathrm{i}\hat{J}_i\theta_i}\,|b\rangle = \sum_j \sum_{a,b} C_{ab}^{(j)}\,\langle a|e^{\mathrm{i}\hat{J}_i\theta_i}\,\hat{J}_-|b\rangle = 0 \tag{6.51}$$

This is a holomorphicity condition and upon using the parametrization (6.44) will be seen to be identical to the condition (6.22), namely, $\mathcal{D}_{\bar{z}}\Psi = 0$. From the group theory point of view, (6.51) means that the state $|b\rangle$ must also be the lowest weight state. If we have a state $|b\rangle$ with $J_3 = -n/2$ and it is also the lowest weight state, then we must have $j = n/2$. Thus only one representation in (6.48) will have nonzero coefficients, identifying the general wave function as

$$\Psi = \sum_a C_{a,-\frac{n}{2}}^{\left(\frac{n}{2}\right)}\,\mathcal{D}_{a,-\frac{n}{2}}^{\left(\frac{n}{2}\right)}(g) \tag{6.52}$$

A general state is a linear combination of $\mathcal{D}_{a,-\frac{n}{2}}^{\left(\frac{n}{2}\right)}(g)$; since a takes $2j+1$ values, we see that the Hilbert space corresponds to a unitary irreducible representation of $SU(2)$ with $j = n/2$. The operators J_i given in (6.29) or (6.30) correspond to the left action on g, i.e.,

$$J_i\,\Psi(g) = \sum_a C_{a,-\frac{n}{2}}^{\left(\frac{n}{2}\right)}\,\mathcal{D}_{a,-\frac{n}{2}}^{\left(\frac{n}{2}\right)}\left(\frac{\sigma_i}{2}\,g\right) = \sum_{a,c} C_{a,-\frac{n}{2}}^{\left(\frac{n}{2}\right)}\,(J_i)_{ac}\,\mathcal{D}_{c,-\frac{n}{2}}^{\left(\frac{n}{2}\right)}(g) \tag{6.53}$$

Here $(J_i)_{ac}$ is the matrix version of $\hat{J}_i$ in the representation with $j = n/2$. We have thus reproduced the previous results from a purely group theoretic point of view.

The explicit construction of the wave functions is also straightforward. Since R_- must annihilate $\Psi(g)$ and we need $R_3\Psi = -(n/2)\Psi$, we see that the wave functions are of the form

$$\Psi(g) = \mathcal{N}\,g_{i_1 2}\,g_{i_2 2}\cdots g_{i_n 2} \tag{6.54}$$

where g_{ij} is given by the 2×2 matrix in (6.44). Once we choose a coordinate patch, we can set θ to some specific value. In particular around $z = 0$, we can set $\theta = 0$. Notice that the second index for each g_{ij} is set to the value 2. It is then easy to see that this does give $-n/2$ for the value of R_3 and that R_- annihilates this since each g_{i2} is in the lowest state for the right translations. In this equation $\mathcal{N}$ is a normalization factor. The wave function Ψ is symmetric in the indices $i_1, i_2, \ldots, i_n$, so that the number of independent components is $n + 1$, as expected. Normalizing these wave functions, we then get the same set as in (6.23), (6.25).

Before we consider the generalization of this to arbitrary groups, it is useful to mention some examples where these results turn up. We may regard the $\Omega = n\omega$ as a $U(1)$ magnetic field which is constant (in the appropriate coordinate frame) on the sphere. If we consider the two-sphere to be embedded in $\mathbb{R}^3$, this constant magnetic field can be viewed as the (radial) field due to a magnetic monopole of charge n at the origin. This identification of the $U(1)$ field is further supported by looking at $R_\pm$. These are translation operators on the sphere, but their commutator is given by $[R_+, R_-]\Psi = 2 R_3 \Psi = -n \Psi$. Since the commutator of covariant derivatives is the field strength in the presence of a gauge field, we can identify a magnetic field for this case as $2 B = n$. (We set the electric charge to be 1; also we took the sphere to have radius equal to 1, otherwise this would read $2 Br^2 = n$ where r is the radius of the sphere.) In the context of charged particle dynamics in the presence of a magnetic monopole, the quantization of the magnetic flux is the Dirac quantization condition. We see that this is equivalent to the quantization of $\int \Omega$. Thus the states (6.52) we find are the angular part of the wave functions for a charged particle in the presence of a magnetic monopole. Also they can be thought of as the lowest Landau levels for a constant magnetic field on the sphere [12, 13]. The left action of the J_i as in (6.53) correspond to the so-called magnetic translations for the Landau levels. So quantum Hall effect on the sphere can be discussed using these wave functions. We will consider some more details of this problem later.

The geometric quantization of the two-sphere can also appear as part of the dynamics of a particle with spin; we get one unitary irreducible representation (UIR) of $SU(2)$, so we have exactly what is needed for spin. It can also be thought of as describing the internal symmetry structures, such as the color degrees of freedom for a particle with nonabelian charges for the case of the color group being $SU(2)$.

6.3 Kähler Spaces of the G/H-Type

The two-sphere $S^2 = SU(2)/U(1)$ is an example of a group coset which is a Kähler manifold. There are many Kähler manifolds which are of the form G/H where H is a subgroup of a compact Lie group G. In particular G/H is a Kähler manifold for any compact Lie group if H is its maximal torus. Another set of Kähler spaces of this type is given by $\mathbb{CP}^k = SU(k + 1)/U(k)$. There are also examples of this type corresponding to noncompact groups. For example, the Lobachevskian space $SL(2, \mathbb{R})/U(1)$ is also a Kähler manifold, although its volume defined by the Kähler two-form is infinite. There are many other cases as well which are interesting from the physics point of view, see later chapters and corresponding references.

In these cases, one can consider theories where the symplectic form is proportional to the Kähler form or is a combination of the generators of $\mathcal{H}^2(M, \mathbb{R})$ for these manifolds and quantize as we have done for the case of S^2. The general result is that they lead to one unitary irreducible representation of the group G, the specific representation being determined by the choice of Ω.

6.3.1 Quantizing $\mathbb{CP}^2$

In most of these cases with G/H structure, it is rather simple and straightforward to construct the Kähler form for these spaces. We will now consider in some detail another example, namely, the quantization of $\mathbb{CP}^2 = SU(3)/U(2)$. A general element of $SU(3)$ can be represented as a unitary (3×3)-matrix. This can be taken to be of the form $g = \exp(it_a \theta^a)$, where the generators $\{t_a\}$ in the 3×3 matrix representation can be chosen as

$$
t_1 = \frac{1}{2}\begin{pmatrix} 0 & 1 & 0 \\ 1 & 0 & 0 \\ 0 & 0 & 0 \end{pmatrix} \quad
t_2 = \frac{1}{2}\begin{pmatrix} 0 & -i & 0 \\ i & 0 & 0 \\ 0 & 0 & 0 \end{pmatrix} \quad
t_3 = \frac{1}{2}\begin{pmatrix} 1 & 0 & 0 \\ 0 & -1 & 0 \\ 0 & 0 & 0 \end{pmatrix}
$$

$$
t_4 = \frac{1}{2}\begin{pmatrix} 0 & 0 & 1 \\ 0 & 0 & 0 \\ 1 & 0 & 0 \end{pmatrix} \quad
t_5 = \frac{1}{2}\begin{pmatrix} 0 & 0 & -i \\ 0 & 0 & 0 \\ i & 0 & 0 \end{pmatrix} \quad
t_6 = \frac{1}{2}\begin{pmatrix} 0 & 0 & 0 \\ 0 & 0 & 1 \\ 0 & 1 & 0 \end{pmatrix} \quad
t_7 = \frac{1}{2}\begin{pmatrix} 0 & 0 & 0 \\ 0 & 0 & -i \\ 0 & i & 0 \end{pmatrix}
$$

$$
t_8 = \frac{1}{\sqrt{12}}\begin{pmatrix} 1 & 0 & 0 \\ 0 & 1 & 0 \\ 0 & 0 & -2 \end{pmatrix} \tag{6.55}
$$

(These matrices t_a form an orthonormal basis for the Lie algebra of $SU(3)$, they are normalized so that $\mathrm{Tr}(t_a t_b) = \frac{1}{2}\delta_{ab}$.) We define a $U(1)$ subgroup by elements of the form $\exp(it_8 \theta^8)$ and we can also define an $SU(2)$ subgroup which commutes with this $U(1)$ subgroup; the latter has elements of the form $U = \exp(it_a \theta^a)$ for $a = 1, 2, 3$. These two subgroups together form the $U(2)$ subgroup of $SU(3)$.[2] Consider now the one-form

$$
\mathcal{A}(g) = i\, w\, \mathrm{Tr}(t_8\, g^{-1} dg) = -i\, w\, \frac{\sqrt{3}}{2}\, u_\alpha\, du_\alpha^* \tag{6.57}
$$

---

[2] Strictly speaking there is an identification of certain elements involved. There is a common $\mathbb{Z}_2$ subgroup for the factors in $SU(2) \times U(1)$ defined by $\mathbb{Z}_2 = \{1, h_Z\}, h_Z = (h_2, h_1)$ with $h_Z^2 = 1$ and

$$
h_2 = \begin{pmatrix} -1_{2\times2} & 0 \\ 0 & 1 \end{pmatrix}, \qquad h_1 = \exp(it_8 \sqrt{12}\,\pi) = \begin{pmatrix} -1_{2\times2} & 0 \\ 0 & 1 \end{pmatrix} \tag{6.56}
$$

The $U(2)$ subgroup is thus given by $SU(2) \times U(1)/\mathbb{Z}_2$.

where g is an element of the group $SU(3)$ and w is a numerical constant; $u_\alpha^* = g_{\alpha 3}$. If h is an element of $U(2) \subset SU(3)$ of the form $h = U \exp(i\, t_8\, \theta)$, we find

$$\mathcal{A}(g\,h) = \mathcal{A}(g) - \frac{w}{2}\, d\theta \tag{6.58}$$

We see that $\mathcal{A}$ changes by a total differential under the $U(2)$-transformations. The two-form $d\mathcal{A}$ is therefore independent of θ or it is invariant under $U(2)$ transformations. Thus the two-form $d\mathcal{A}$ is defined on the coset space $SU(3)/U(2)$. Evidently it is closed ($dd\mathcal{A} = 0$ since $d^2 = 0$), but it is not exact since the corresponding one-form $\mathcal{A}$ is not globally defined on $SU(3)/U(2)$, but only on $G = SU(3)$. Thus $d\mathcal{A}$ is a nontrivial element of $\mathcal{H}^2(SU(3)/U(2), \mathbb{R}) = \mathcal{H}^2(\mathbb{CP}^2, \mathbb{R})$. There will be quantization conditions on w and the lowest possible choice, with our choice of normalization for t_8, will be $2/\sqrt{3}$. With this choice, we define the Kähler 2-form for $SU(3)/U(2)$ as

$$\omega = d\left(i\frac{2}{\sqrt{3}}\, \mathrm{Tr}(t_8\, g^{-1} dg)\right) \tag{6.59}$$

The connection with the complex projective space is clarified by introducing $Z_\alpha = \rho\, u_\alpha$, where ρ is an arbitrary complex number, not equal to zero. We can then consider

$$\mathcal{A} = -i\, w\, \frac{\sqrt{3}}{2}\, \frac{Z \cdot d\bar{Z}}{\bar{Z} \cdot Z} = -i\, w\, \frac{\sqrt{3}}{2}\, \left[u_\alpha\, du_\alpha^* + d\log\rho\right] \tag{6.60}$$

This $\mathcal{A}$ differs from (6.57) by a total derivative and hence $d\,\mathcal{A}$ will be the same for both $\mathcal{A}$'s. We thus see that we can write ω as

$$\omega = -i\left[\frac{dZ \cdot d\bar{Z}}{(Z \cdot \bar{Z})} - \frac{dZ \cdot \bar{Z}\, Z \cdot d\bar{Z}}{(Z \cdot \bar{Z})^2}\right] \tag{6.61}$$

which is the expected Kähler form on $\mathbb{CP}^2$.

As the symplectic form for quantization of the phase space $\mathbb{CP}^2$, we can consider any integral multiple of ω; we need an integral multiple, since the integrals of $\Omega = d\mathcal{A}$ over nontrivial two-cycles on $\mathbb{CP}^2$ will have to be integers. Thus the possible choices for w are of the form $w = 2\,n/\sqrt{3}$, $n \in \mathbb{Z}$.[3] Therefore we will consider the symplectic two-form

$$\Omega = -i\frac{2\,n}{\sqrt{3}}\, \mathrm{Tr}\left(t_8\, g^{-1} dg \wedge g^{-1} dg\right)$$

$$= n\,\omega = -i\,n\left[\frac{dZ \cdot d\bar{Z}}{(Z \cdot \bar{Z})} - \frac{dZ \cdot \bar{Z}\, Z \cdot d\bar{Z}}{(Z \cdot \bar{Z})^2}\right] \tag{6.62}$$

---

[3] It should be kept in mind that different choices of w correspond to different theories and different physics.

The action which leads to the chosen $\mathcal{A}$ and the Ω in (6.62) is

$$S = i\frac{2n}{\sqrt{3}} \int dt \; \mathrm{Tr}(t_8 \, g^{-1}\dot{g}) \tag{6.63}$$

Again, for e^{iS} to be well defined on $\mathbb{CP}^2$, the values of w will have to be restricted to the form given above, namely, $w = 2n/\sqrt{3}, n \in \mathbb{Z}$. The wave functions are functions on $SU(3)$ subject to the restrictions given by the action of $SU(2)$ and $U(1)$ and a holomorphicity condition, which is the polarization condition. In other words, we can write, using the Wigner $\mathcal{D}$-functions for $SU(3)$,

$$\Psi \sim \mathcal{D}^{(r)}_{AB}(g) = \langle r, A|\, \hat{g}\, |r, B \rangle \tag{6.64}$$

Here (r) is a set of indices which labels the representation, A, B label the states within a representation. Only the finite-dimensional (and hence unitary) representations can occur here, since they form a complete set for functions on $SU(3)$.

The groups involved in the quotient can be taken as the right action on g. The transformation law for $\mathcal{A}$ then tells us that Ψ must transform as

$$\Psi(g\,h) = \Psi(g) \; \exp\left(-i\frac{n}{\sqrt{3}}\,\theta\right) \tag{6.65}$$

This shows that the wave functions must be invariant, i.e., singlets, under the $SU(2)$ subgroup and carry a definite charge $n/\sqrt{3}$ under the $U(1)$ subgroup generated by t_8. This restricts the choice of values for the state $|r, B\rangle$ in (6.64). Further $w = 2n/\sqrt{3}$ must also be quantized so that it can be one of the allowed values in the representations of $SU(3)$ in (6.64). This is the same as what we already found, namely, that n must be an integer.

One has to choose a polarization condition as well. The generators of $SU(3)$ can be divided into those of the $SU(2)$ and $U(1)$ subgroups, and the coset ones which correspond to t_i with $i = 4, 5, 6, 7$. These can be grouped into t_a, $t_{\bar{a}}$, with $a, \bar{a} = 1, 2$, corresponding to $t_1 = t_4 + it_5, t_2 = t_6 + it_7$, and their conjugates. Correspondingly, we can define the right translation operators

$$R_a \, g = g\, t_a, \qquad R_{\bar{a}} \, g = g\, t_{\bar{a}} \tag{6.66}$$

As the holomorhic polarization condition, we choose

$$R_{\bar{a}} \, \Psi(g) = 0 \tag{6.67}$$

This requires the state $|r, B\rangle$ to be a lowest weight state. This requirement, along with the earlier statement that $|r, B\rangle$ should be an $SU(2)$ singlet with eigenvalue $-n/\sqrt{3}$ for the t_8-transformation, completely fixes the representation r and the state $|r, B\rangle$. However, the left index A in (6.64) is free, taking values corresponding to the possible states in the representation r. Thus the result of the quantization is to yield a Hilbert space which is one unitary irreducible representation of the group $SU(3)$.

We can also carry out an explicit construction of the wave functions which form a basis of the Hilbert space $\mathcal{H}$ along the lines of how it was done for $S^2 = SU(2)/U(1)$ in (6.54). They are of the form

$$\Psi(g) = \mathcal{N} \, g_{\alpha_1 3} \, g_{\alpha_2 3} \cdots g_{\alpha_n 3} \tag{6.68}$$

where $g_{\alpha\beta} \in SU(3)$ is the 3×3 $SU(3)$ matrix in the fundamental representation. Since $(gt_8)_{\alpha 3} = -(1/\sqrt{3})g_{\alpha 3}$ and $R_{\bar{a}}g_{\alpha 3} = 0$, we see that the requirements (6.65) and (6.67) are indeed satisfied by $\Psi(g)$ in (6.68). From the left action of $SU(3)$ on $g_{\alpha 3}$'s, we see that Ψ is in the symmetric rank-n representation of $SU(3)$. This is the UIR obtained for the choice of $\Omega = n\,\omega$ or the action in (6.63). The dimension of the Hilbert space is seen to be $N = \frac{1}{2}(n+1)(n+2)$.

The volume element for $\mathbb{CP}^2$ is defined by the Kähler form ω as $\omega \wedge \omega$. We will use a normalized volume where the total volume of $\mathbb{CP}^2$ is taken to be 1. In terms of local coordinates z_i, $i = 1, 2$, with $g_{i3} = z_i/\sqrt{1 + \bar{z} \cdot z}$, the volume is given by

$$d\mu = \frac{2}{\pi^2} \frac{d^2 z\, d^2 \bar{z}}{(1 + \bar{z} \cdot z)^3} \tag{6.69}$$

The normalized version of Ψ's in (6.68) is thus

$$\Psi(g) = \sqrt{N} \sqrt{\frac{n!}{k_1! k_2! (n-s)!}} \frac{z_1^{k_1} z_2^{k_2}}{(1 + \bar{z} \cdot z)^{\frac{n}{2}}} \tag{6.70}$$

$$k_1, k_2 = 1, 2, \ldots, n, \qquad s = k_1 + k_2$$

6.3.2 Quantizing General G/H Spaces

More generally, with a view of obtaining UIRs of a compact Lie group G, one can take

$$\mathcal{A}(g) = i \sum_a w_a \text{Tr}(t_a \, g^{-1} dg) \tag{6.71}$$

where t_a are diagonal elements of the Lie algebra of G (in some suitable orthonormal basis) and w_a are a set of numbers. The number of independent generators t_a is given by the rank of the group. The elements of the group given by the exponential map $e^{it_a \theta^a}$ form the subgroup T which is the maximal torus of G.

Notice that, under $g \rightarrow gh$, $\mathcal{A}$ will change by a total differential if h commutes with the factor $\sum_a w_a t_a$ in (6.71). The subgroup of G consisting of all elements which commute with $\sum_a w_a t_a$ is the subgroup H. While $\mathcal{A}$ changes under right translations by h, $d\mathcal{A}$ is invariant, and so $d\mathcal{A}$ will be a closed nonexact form on G/H.

If some of the eigenvalues of $\sum_a w_a t^a$ are equal, H can be larger than the maximal torus. Upon quantization, for suitably chosen w_a, we will get one unitary irreducible representation of G and w_a will be related to the highest weights defining the representation.[4]

There is another way to think about this problem. Let us say that we want to construct a unitary irreducible representation (UIR) of a group G. We ask the question: Is there a classical action which upon quantization gives exactly one UIR of the group G? Recall that if we quantize the rigid rotor we get all UIR's of the angular momentum group $SO(3)$. That is not what we want, we want one and only one representation. The answer to this is the action

$$\mathcal{S} = i \sum_a w_a \int dt \, \mathrm{Tr}(t_a \, g^{-1} \dot{g}) \tag{6.72}$$

with the choice of $\{w_a\}$ determined by which representation we wish to obtain upon quantization. This action is known as the co-adjoint orbit action, since it is defined on the orbit $g w_a t_a \, g^{-1}$ of the group G. Often it is also referred to as the Kostant-Kirillov-Souriau action.

One can use an action similar to (6.72) for noncompact groups as well. The key here is that, since we are quantizing the system, the representation we obtain is unitary. Thus if one carries out the quantization of $SL(2, \mathbb{R})/U(1)$, we will get a UIR of $SL(2, \mathbb{R})$. Such representations are infinite dimensional since $SL(2, \mathbb{R})$ is noncompact. The representation obtained will be one of the series needed for the completeness relation for functions on $SL(2, \mathbb{R})$. The infinite dimensionality is also in agreement with the semiclassical counting of the dimension of the Hilbert space since the phase volume (= the volume of $SL(2, \mathbb{R})/U(1)$ as measured by its Kähler form) is infinite.

6.3.3   A Note on an Index Theorem

It is interesting to see the dimension of the Hilbert space in another way [14]. The polarization conditions (6.22) and (6.67) express the $\bar{\partial}$-closure of Ψ with a $U(1)$ gauge field $\mathcal{A}$ and on a space of Riemannian curvature R_{12}. Generally, the number of normalizable zero modes of the $\bar{\partial}$ operator for a vector bundle V can be obtained using the index theorem for the twisted Dolbeault complex. Explicitly, this is expressed as

$$\mathrm{Index}(\bar{\partial}_V) = \int_M \mathrm{td}(M) \wedge \mathrm{ch}(V) \tag{6.73}$$

---

[4] We use the term highest weight in the general algebraic sense, although the actual eigenvalue may be the lowest value in the representation as in some of the previous examples.

where td is the Todd class of the complex tangent space of the manifold and $\text{ch}(V) = \text{Tr}(e^{F/2\pi})$ is the Chern character. (F is the curvature of the vector bundle.) The Todd class itself can be expressed in terms of the Chern classes, which, for a vector bundle with curvature $\mathcal{F}$, are given by

$$\det\left(1 + \frac{i\mathcal{F}}{2\pi} t\right) = \sum_i c_i t^i \tag{6.74}$$

The Todd class may also be represented, via the splitting principle, in terms of a generating function as

$$\text{td} = \prod_i \frac{x_i}{1 - e^{-x_i}} \tag{6.75}$$

where x_i represent the "eigenvalues" of the curvature in a suitable canonical form (diagonal or the canonical antisymmetric form for real antisymmetric $i\mathcal{F}$). Explicitly, the Todd class, up to the 3-form level, is given as

$$\text{td} = 1 + \frac{1}{2}c_1 + \frac{1}{12}(c_1^2 + c_2) + \frac{1}{24}c_1 c_2 + \cdots \tag{6.76}$$

The curvature $i\mathcal{F}$ relevant for us will be the curvature of the complex tangent space of the manifold, $T_c M$. After multiplying out $\text{td}(M)$ and $\text{ch}(V)$, the differential form of the dimension appropriate to the space of interest should be used as the integrand in (6.73).

In the two-dimensional case, the Todd class $\text{td}(M)$ is $R/4\pi$ and the Chern character $\text{ch}(V) = \text{Tr}(e^{F/2\pi})$ is $\Omega/2\pi$ for us. The number of normalizable solutions to (6.22) for the two-sphere is thus

$$\text{Index}(\bar{\partial}_V) = \int_M \frac{\Omega}{2\pi} + \int_M \frac{R}{4\pi} = n + 1 \tag{6.77}$$

Notice that, semiclassically, we should expect the number of states to be $\int \Omega/2\pi = n$. The extra one comes from the Euler number in this case. (The semiclassical counting is supposed to apply only for large n, so this is all consistent with expectations.)

For the case of $\mathbb{CP}^2$, we have

$$\text{Index}(\bar{\partial}_V) = \int_M \left[\frac{1}{2}\text{Tr}\left(\frac{F}{2\pi}\right)^2 + \frac{1}{2}c_1\text{Tr}\left(\frac{F}{2\pi}\right) + \frac{1}{12}(c_1^2 + c_2)\right]$$

$$= \int_M \left[\frac{1}{2}\text{Tr}\left(\frac{\Omega}{2\pi}\right)^2 + \frac{1}{2}\text{Tr}\left(\frac{R}{2\pi}\right)\text{Tr}\left(\frac{\Omega}{2\pi}\right)\right]$$

$$+ \frac{1}{12}\int_M \left[\frac{3}{2}\left(\text{Tr}\frac{R}{2\pi}\right)^2 - \frac{1}{2}\text{Tr}\left(\frac{R}{2\pi}\frac{R}{2\pi}\right)\right] \tag{6.78}$$

where $\Omega = n\omega$. Further, in terms of the Kähler two-form,

$$\text{Tr}\frac{R}{2\pi} = 3\,\frac{\omega}{2\pi}, \qquad \text{Tr}\left(\frac{R}{2\pi}\frac{R}{2\pi}\right) = 3\left(\frac{\omega}{2\pi}\right)^2 \tag{6.79}$$

With our chosen normalization for the total volume of $\mathbb{CP}^2$,

$$\int_M \left(\frac{\omega}{2\pi}\right)^2 = 1 \tag{6.80}$$

The index is then easily evaluated as

$$\text{Index}(\bar{\partial}_V)\Big|_{\mathbb{CP}^2} = \frac{1}{2}(n+1)(n+2) \tag{6.81}$$

This agrees with what we obtained as the dimension of the Hilbert space with the explicit construction of the wave functions as in (6.68).

6.3.4 A Short Historical Note

Historically, geometric quantization arose out of representation theory for groups. The construction of UIR's of a compact group using the Kähler two-form on G/T where T is the maximal torus was carried out in the 1950s. It goes by the name of Borel-Weil-Bott theory. Geometric quantization was developed in the 1970s (by Kostant, Souriau, Kirillov and others) as an attempt to generalize this to arbitrary symplectic manifolds. The use of actions of the form (6.72) for various physical problems was pursued in the 1970s by Wong, Balachandran and others [10, 11]. This action (6.72) may also be viewed as the prototypical Wess-Zumino term. The usual Wess-Zumino term was introduced in the context of meson physics by Wess and Zumino in 1971 as an effective action for anomalies [15]. It was developed and its full import was realized in the work of Witten [16]. (In this context, Novikov's work on the Wess-Zumino term in a (2+1)-dimensional setting should be mentioned, although the physics implications were not fully evident [17]. There were also a few other earlier papers which focused on certain aspects of the Wess-Zumino term.) Also, as mentioned before, geometric quantization applies to the quantum Hall effect as well, both in two dimensions and in higher dimensions [12, 13].

Problems

6.1 Find the spin connection and curvature and its integral for S^2.

6.2 Carry out the infinitesimal transformations generated by the vector fields $\xi_\pm$, ξ_3 given in (6.27) and show that they are isometries of S^2.

6.3 Identify the generators of left translations of g in Ω of (6.47).

6.4 Consider the geometric quantization of the hyperboloidal space $SL(2, \mathbb{R})/U(1)$. The canonical two-form is given by

$$\Omega = 2i\lambda \frac{dz \wedge d\bar{z}}{(1 - z\bar{z})^2}$$

This applies to the region $z\bar{z} \leq 1$. Show that $V_+ = iz^2 \partial_z - i\partial_{\bar{z}}$, $V_- = -i\bar{z}^2 \partial_{\bar{z}} + i\partial_z$, $V_3 = iz\partial_z - i\bar{z}\partial_{\bar{z}}$ are Hamiltonian vector fields. Identify the nature of the wave functions, the inner product in the holomorphic polarization and the operators corresponding to $V_\pm$, V_3.

6.5 Consider the geometric quantization of the hyperboloidal space $SL(2, \mathbb{R})/U(1)$. The canonical two-form is given by

$$\Omega = 2i\lambda \frac{dz \wedge d\bar{z}}{(1 - z\bar{z})^2}$$

This applies to the region $z\bar{z} \leq 1$. Show that $V_+ = iz^2 \partial_z - i\partial_{\bar{z}}$, $V_- = -i\bar{z}^2 \partial_{\bar{z}} + i\partial_z$, $V_3 = iz\partial_z - i\bar{z}\partial_{\bar{z}}$ are Hamiltonian vector fields. Identify the nature of the wave functions, the inner product in the holomorphic polarization and the operators corresponding to $V_\pm$, V_3.

Open Access This chapter is licensed under the terms of the Creative Commons Attribution 4.0 International License (http://creativecommons.org/licenses/by/4.0/), which permits use, sharing, adaptation, distribution and reproduction in any medium or format, as long as you give appropriate credit to the original author(s) and the source, provide a link to the Creative Commons license and indicate if changes were made.

The images or other third party material in this chapter are included in the chapter's Creative Commons license, unless indicated otherwise in a credit line to the material. If material is not included in the chapter's Creative Commons license and your intended use is not permitted by statutory regulation or exceeds the permitted use, you will need to obtain permission directly from the copyright holder.

Chapter 7
The Chern-Simons Theory in 2+1 Dimensions

The Chern-Simons (CS) theory is a gauge theory in two space (and one time) dimensions [18, 19]. The action is given by

$$S = -\frac{k}{4\pi} \int_{\Sigma \times [t_i, t_f]} \mathrm{Tr} \left[A \, \mathrm{d}A + \frac{2}{3} A A A \right]$$
$$= -\frac{k}{4\pi} \int_{\Sigma \times [t_i, t_f]} \mathrm{d}^3 x \, \epsilon^{\mu\nu\alpha} \, \mathrm{Tr} \left[A_\mu \partial_\nu A_\alpha + \frac{2}{3} A_\mu A_\nu A_\alpha \right] \qquad (7.1)$$

Here A_μ is the Lie algebra-valued gauge potential, $A_\mu = -i \, t_a A_\mu^a$, corresponding to a compact Lie group G. t_a are hermitian matrices forming a basis of the Lie algebra in the fundamental representation of the gauge group. We shall take the gauge group to be $G = SU(N)$ in what follows and, as before, normalize $\{t_a\}$ to obey the condition $\mathrm{Tr}(t_a t_b) = \frac{1}{2}\delta_{ab}$, with the structure constants f_{abc} defined by $[t_a, t_b] = \mathrm{i} f_{abc} t_c$. Thus, for example, for the case of the gauge group being $SU(3)$, the set of matrices t_a can be taken as the ones given in (6.55). In addition to the choice of the group, the theory has one parameter k, which is referred to as the *level number* of the Chern-Simons form. It is a real constant whose precise value we do not need to specify at this stage. The two-dimensional spatial manifold is denoted by Σ. The classical equations of motion for the theory are

$$F_{\mu\nu} = 0 \qquad (7.2)$$

We will be interested in the case of Σ being a Riemann surface, the complex structure on Σ facilitating geometric quantization with holomorphic polarization. However, it is useful to consider some features of quantization on $\Sigma = \mathbb{R}^2$ first. We may think of $\mathbb{R}^2$ as a disc of radius r, with $r \to \infty$ eventually. (On a compact manifold without boundary, the action is invariant under gauge transformations; we will take $F_{\mu\nu}$ to vanish as $r \to \infty$ so that we have gauge invariance for $\Sigma = \mathbb{R}^2$ as

© The Author(s) 2024
V. P. Nair, *Geometric Quantization and Applications to Fields and Fluids*,
SpringerBriefs in Physics, https://doi.org/10.1007/978-3-031-65801-3_7

well.) Consider now the gauge $A_0 = 0$. The term involving the time-derivatives in the action is

$$S = -\frac{k}{8\pi} \int \epsilon^{ij} A_i^a \partial_0 A_j^a + \cdots \tag{7.3}$$

The surface term arising from the time-integration in the variation of the action is

$$\delta S = -\frac{k}{8\pi} \int_\Sigma \epsilon^{ij} A_i^a \delta A_j^a \Big|_{t_i}^{t_f} + \cdots \tag{7.4}$$

From this we can read off the symplectic potential and the corresponding canonical two-form as

$$\mathcal{A} = -\frac{k}{8\pi} \int_\Sigma \epsilon^{ij} A_i^a \delta A_j^a + \delta\rho[A]$$

$$\Omega = -\frac{k}{8\pi} \int_\Sigma \epsilon^{ij} \delta A_i^a \delta A_j^a \tag{7.5}$$

where ρ is arbitrary functional of A. Its presence in $\mathcal{A}$ indicates the freedom of canonical transformations. The Hamiltonian vector field corresponding to A_j^a is thus

$$V_{A_j^a} = -\frac{4\pi}{k} \epsilon_{jk} \frac{\delta}{\delta A_k^a} \tag{7.6}$$

The basic Poisson bracket is thus given by

$$\{A_i^a(x), A_j^b(y)\} = \frac{4\pi}{k} \epsilon_{ij} \delta^{ab} \delta^{(2)}(x - y) \tag{7.7}$$

The phase space of the theory is the space of the fields $A_i^a(x)$, i.e., the space of Lie algebra-valued one-forms on $\mathbb{R}^2$. We will assume that these obey some (mild) conditions which ensure finiteness of the action or $\int_\Sigma F^2$.

A gauge transformation of the field A is given by

$$A \rightarrow A^g = g A g^{-1} - dg\, g^{-1}, \qquad g \in G \tag{7.8}$$

For infinitesimal transformations, $g \approx 1 - i t_a \theta^a$, and we get

$$\delta_g A_i^a = -\left(\partial_i \theta^a - f^{cab} A_i^c \theta^b\right) = -(D_i \theta)^a \tag{7.9}$$

The basic commutation rule which follows from the Poisson bracket (7.7) is

$$\left[A_i^a(x), A_j^b(y)\right] = \frac{4\pi i}{k} \epsilon_{ij} \delta^{ab} \delta^{(2)}(x - y) \tag{7.10}$$

Using this relation we can directly check that

$$\left[G_0(\theta), A_i^a(x)\right] = \mathrm{i}(D_i\theta)^a = -\mathrm{i}\,\delta A_i^a \tag{7.11}$$

where

$$G_0(\theta) = \frac{k}{8\pi} \int_\Sigma \theta^a \, \epsilon^{kl} \, F_{kl}^a \tag{7.12}$$

We see that $G_0(\theta)$ generates the transformation (7.9). Although $G_0(\theta)$ does generate gauge transformations, the nature of the functions $\theta^a(x)$ is important in obtaining a consistent action on functionals of A. This can be illustrated by a simple example of how $G_0(\theta)$ acts on $\int A^2$. Again using the commutation rule (7.10), we find

$$\left[G_0(\theta), \int A^2\right] = 2\mathrm{i} \int_{x,y} \left[-\theta^a(y)\frac{\partial}{\partial y^i}\delta^{(2)}(y-x)A_i^a(x)\right.$$
$$\left. -f^{cab} A_i^c(y)\theta^b(y)A_i^a(x)\delta^{(2)}(y-x)\right] \tag{7.13}$$

We can simplify this expression in two ways. If we integrate over x first, we get

$$\left[G_0(\theta), \int A^2\right] = -2\mathrm{i} \int_\Sigma \theta^a \, \partial \cdot A^a \tag{7.14}$$

We can also write $\partial_y\delta^{(2)}(y-x) = -\partial_x\delta^{(2)}(y-x)$ and integrate over y first. This gives the expression

$$\left[G_0(\theta), \int A^2\right] = 2\mathrm{i} \int_\Sigma \partial_i\theta^a A_i^a$$
$$= 2\mathrm{i}\left[-\int_\Sigma \theta^a\partial \cdot A^a + \oint_{\partial\Sigma} \theta^a A^a \cdot \mathrm{d}S\right] \tag{7.15}$$

Comparing (7.14) and (7.15), we see that the consistent evaluation of the action of $G_0(\theta)$ on functionals of A will require the functions θ^a to vanish on the boundary, or as $r \to \infty$. The subscript on $G_0(\theta)$ was introduced in anticipation of this to signify that the functions θ must vanish at the boundary.

We can now consider reducing the theory to the physical degrees of freedom which are gauge-invariant by starting with wave functions which are arbitrary functionals of A and then selecting the physical states by imposing the condition

$$G_0(\theta)\,\Psi = 0 \tag{7.16}$$

This condition is just the Gauss law of the Chern-Simons theory. Notice that it is also one of the equations of motion in (7.2). The equation $F_{ij} = 0$ cannot be a Heisenberg equation of motion, since it only involves spatial derivatives of the data at a fixed time, namely A_i^a. Therefore it must be viewed as a condition restricting the phase space variables to the physical subspace. The condition (7.16) is one way this restriction can be implemented in the quantum theory.

The analysis given above also clarifies the nature of the space of fields. Let

$$\mathfrak{F} = \{\text{Set of all gauge potentials } A_i\} \tag{7.17}$$

where, as stated before, the potential A_i is a Lie algebra-valued one-form on Σ, obeying some finiteness condition on integrals like the action or $\int_\Sigma F^2$. This is the space of fields we start with; it is also the starting phase space for the Chern-Simons theory with Ω as in (7.5). We also define the space of gauge transformations as

$$\mathfrak{G}_* = \{g(x) : \mathbb{R}^2 \to G, \text{ with } g \to 1 \text{ as } r \to \infty\} \tag{7.18}$$

This is the relevant space as seen from our discussion of the boundary value of θ^a in $G_0(\theta)$. It is obviously a group under pointwise multiplication. The physical space of fields is then given by $\mathfrak{C} = \mathfrak{F}/\mathfrak{G}_*$. If we now consider a one-point compactification of $\mathbb{R}^2$ to S^2, with the point at infinity mapped to a point x_0 on the sphere, the definition of $\mathfrak{G}_*$ becomes

$$\mathfrak{G}_* = \{g(x) : S^2 \to G, \text{ with } g \to 1 \text{ at } x_0\} \tag{7.19}$$

Such maps from a compact space to G, with g set to a fixed value (in our case $g = 1$) at a specific point x_0 are called pointed maps. These are the relevant ones to be factored out in the quantization of the gauge theory [20].

The gauge transforms of a given gauge field or connection A, the set of fields $gAg^{-1} - dgg^{-1}$ for all $g \in \mathfrak{G}_*$ define the orbit of A. Since these are all identified in $\mathfrak{C} = \mathfrak{F}/\mathfrak{G}_*$, points in $\mathfrak{C}$ correspond to gauge orbits and $\mathfrak{C}$ is often referred to as the gauge-orbit space. Also, wave functions are sections of a line bundle on the gauge-orbit space $\mathfrak{C}$.

We close this section with a couple of remarks. Analogous to $G_0(\theta)$ we can define the operator

$$G(\varphi) = \frac{k}{8\pi} \int_\Sigma \epsilon^{kl} \left[-2 \, \partial_k \varphi^a A_l^a + f^{abc} \varphi^a A_k^b A_l^c \right] \tag{7.20}$$

We do not assume that φ^a vanishes as $r \to \infty$. It is easy to verify that there is no subtlety in the action of $G(\varphi)$ on functionals of A. For example,

$$\left[G(\varphi), \int A^2 \right] = 2\mathrm{i} \int \partial \varphi^a \cdot A \tag{7.21}$$

Further, we can check that

$$[G_0(\theta), G(\varphi)] = \mathrm{i} \, G_0(\theta \times \varphi), \qquad (\theta \times \varphi)^a = f^{abc} \theta^b \varphi^c \tag{7.22}$$

Notice that, since θ^a vanishes, $(\theta \times \varphi)^a \to 0$ as $r \to \infty$. Therefore the expression on the right hand side of (7.22) can be understood as $G_0(\theta \times \varphi)$. An important consequence of this equation is that, if Ψ is a physical state obeying the Gauss law, i.e., satisfying (7.16),

$$G_0(\theta)\Big[G(\varphi)\Psi\Big] = G(\varphi)G_0(\theta)\Psi + i\,G_0(\theta \times \varphi)\Psi = 0 \qquad (7.23)$$

This shows that $G(\varphi)\Psi$ is a physical state. By different choices of φ^a, the action of $G(\varphi)$ will generate a number of new physical states from Ψ. Even though the data defining such states is φ^a, only the values of φ^a as $r \to \infty$ are relevant since the values at finite r can be modified by the action of $G_0(\theta)$. So the states generated by $G(\varphi)$ are called the edge modes of the theory. They represent physical degrees of freedom residing on the boundary or at $r \to \infty$. (For more details on edge states for the Chern-Simons theory, see [18, 21, 22].)

Our second remark is about the gauge-orbit space $\mathfrak{C}$. Consider the gauge transform of a connection A given by $A^g = gAg^{-1} - dgg^{-1}$. If this is equal to A itself, i.e., if

$$A^g = gAg^{-1} - dgg^{-1} = A \qquad (7.24)$$

for a nontrivial $g(x) \in \mathfrak{G}_*$, i.e., for $g \neq 1$, we say that the connection is reducible. When we make the identification of gauge transforms of A to go from $\mathfrak{F}$ to $\mathfrak{C} = \mathfrak{F}/\mathfrak{G}_*$, such connections can lead to singularities. Pointed maps however can avoid the reducible connections for a manifold like S^2. Consider solving (7.24) for g for a given A. Towards this first consider the Wilson line associated to a connection A. It is defined by

$$W_C(x, y, A) = \mathcal{P}\exp\left(-\int_{y,C}^{x} A\right) \qquad (7.25)$$

where $\mathcal{P}$ denotes path-ordering. In general, $W_C(x, y, A)$ depends on a curve C connecting the points y and x and it is also defined by the chosen A. Since $W_C(x, y, A)$ obeys the equation

$$n \cdot \big[\partial_x + A(x)\big]W_C(x, y, A) = 0$$
$$n \cdot \big[\partial_y W_C(x, y, A) - W_C(x, y, A)A(y)\big] = 0 \qquad (7.26)$$

where n_i denotes the tangent to the curve, it is easy to verify that

$$W_C(x, y, A^g) = g(x)\, W_C(x, y, A)\, g^{-1}(y) \qquad (7.27)$$

Thus for a reducible connection, we get

$$g(x)\, W_C(x, y, A)\, g^{-1}(y) = W_C(x, y, A) \qquad (7.28)$$

Notice also that by reversing the curve, we get the inverse to $W(x, y, A)$ Using these results, we can write $g(x)$ at any point for a reducible connection as

$$g(x) = W_C(x, y, A)\, g(y)\, W_C^{-1}(x, y, A) \tag{7.29}$$

This gives $g(x)$ at any point x, for any g which preserves A as in (7.24), in terms of g at the point y. If we choose $y = x_0$, which is the point where we set $g = 1$ for pointed maps, we see that this equation implies $g(x) = 1$ for all x. Thus there is no nontrivial $g(x)$ consistent with setting $g(x_0) = 1$. The factoring out of $\mathfrak{G}_*$ can be carried out without encountering singularities. (If the group has a center, then we can set $g(x_0)$ to be an element of the center, and a similar reduction can be done with $g(x)$ equal to the same element of the center, since such an element will commute with W_C. The theory then has sectors labeled by the elements of the center.)

7.1 Analysis on $S^2 \times \mathbb{R}$

We now turn to the more detailed analysis of Chern-Simons theory on $S^2 \times \mathbb{R}$ [23, 24]. We shall use complex coordinates for $\Sigma = S^2$. In terms of local Cartesian components, the complex gauge fields are $A_z = \frac{1}{2}(A_1 + iA_2)$, $A_{\bar{z}} = \frac{1}{2}(A_1 - iA_2)$. In terms of the complex components, the symplectic two-form Ω is given by

$$\begin{aligned}
\Omega &= -\frac{ik}{\pi} \int_\Sigma d\mu_\Sigma \, \mathrm{Tr}\big(\delta A_{\bar{z}} \delta A_z\big) \\
&= \frac{ik}{2\pi} \int_\Sigma d\mu_\Sigma \, \delta A_{\bar{z}}^a \delta A_z^a
\end{aligned} \tag{7.30}$$

The complex structure on Σ induces a complex structure on $\mathfrak{F}$. We may take A_z, $A_{\bar{z}}$ as the local complex coordinates on $\mathfrak{F}$. In fact, we have a Kähler structure on $\mathfrak{F}$, Ω being the Kähler two-form with the Kähler potential

$$K = \frac{k}{2\pi} \int_\Sigma A_{\bar{z}}^a A_z^a \tag{7.31}$$

The Hamiltonian vector fields corresponding to A_z and $A_{\bar{z}}$ are

$$A_z^a(z) \longrightarrow -\frac{2\pi}{ik} \frac{\delta}{\delta A_{\bar{z}}^a}, \qquad A_{\bar{z}}^a(z) \longrightarrow \frac{2\pi}{ik} \frac{\delta}{\delta A_z^a} \tag{7.32}$$

The Poisson brackets for $A_{\bar{z}}$, A_z are obtained using the general formula (2.11) as

$$\{A_z^a(z), A_w^b(w)\} = 0$$
$$\{A_{\bar{z}}^a(z), A_{\bar{w}}^b(w)\} = 0$$
$$\{A_z^a(z), A_{\bar{w}}^b(w)\} = -\frac{2\pi i}{k} \delta^{ab} \delta^{(2)}(z - w) \tag{7.33}$$

These become commutation rules upon quantization.

The infinitesimal version of the gauge transformations (7.8) (for $g \approx 1 - it_a \theta^a$) corresponds to the vector field

$$\xi = -\int_\Sigma \left[(D_z \theta)^a \frac{\delta}{\delta A_z^a} + (D_{\bar{z}} \theta)^a \frac{\delta}{\delta A_{\bar{z}}^a} \right] \tag{7.34}$$

where D_z and $D_{\bar{z}}$ denote the corresponding gauge covariant derivatives. By contracting this with Ω we get

$$i_\xi \Omega = -\delta \left[\frac{ik}{2\pi} \int_\Sigma F_{z\bar{z}}^a \theta^a \right] \tag{7.35}$$

This identifies the generator of infinitesimal gauge transformations is

$$G_0(\theta) = \frac{ik}{2\pi} \int_\Sigma \theta^a \, F_{z\bar{z}}^a \tag{7.36}$$

This is same as (7.12) written using complex components. Notice also that, for finite transformations, we get

$$\Omega(A^g) - \Omega(A) = \delta \left[\frac{ik}{\pi} \int_\Sigma \mathrm{Tr}(g^{-1}\delta g \, F_{z\bar{z}}) \right]$$

$$= \delta \left[\frac{k}{2\pi} \int_\Sigma \mathrm{Tr}(g^{-1}\delta g \, F) \right] \tag{7.37}$$

(F in the second line of this equation is the two-form $dA + A\,A$.)

The construction of the wave functions proceeds as follows. One has to consider a line bundle on the phase space with curvature Ω. Sections of this bundle give the prequantum Hilbert space. In other words we consider functionals $\Phi[A_z, A_{\bar{z}}]$ with the condition that under the canonical transformation $A \to A + \delta\Lambda$, $\Phi \to e^{i\Lambda} \Phi$. The inner product on the prequantum Hilbert space is given by

$$\langle 1|2 \rangle = \int d\mu(A_z, A_{\bar{z}}) \, \Phi_1^*[A_z, A_{\bar{z}}] \, \Phi_2[A_z, A_{\bar{z}}] \tag{7.38}$$

where $d\mu(A_z, A_{\bar{z}})$ is the Liouville measure associated with Ω. Given the Kähler structure Ω, this is just the volume $[dA_z dA_{\bar{z}}]$ associated with the metric $\|\delta A\|^2 = \int_\Sigma \delta A_{\bar{z}}^a \delta A_z^a$.

The wave functions so constructed depend on all phase space variables. We must now choose the polarization conditions on the Φ's so that they depend only on half the number of phase space variables, leading to the reduction of the prequantum Hilbert space to the Hilbert space of the quantum theory. Given the Kähler structure of the phase space, the most appropriate choice is the Bargmann polarization. With a specific choice of $\rho[A]$ in (7.5), the symplectic potential can be taken as

$$\mathcal{A} = -\frac{ik}{2\pi} \int_\Sigma \mathrm{Tr}\big(A_{\bar{z}}\delta A_z - A_z\delta A_{\bar{z}}\big) = \frac{ik}{4\pi} \int_\Sigma \big(A_{\bar{z}}^a\delta A_z^a - A_z^a\delta A_{\bar{z}}^a\big) \quad (7.39)$$

The covariant (functional) derivatives with $\mathcal{A}$ as the potential are

$$\nabla = \big(\frac{\delta}{\delta A_z^a} + \frac{k}{4\pi}A_{\bar{z}}^a\big), \qquad \overline{\nabla} = \big(\frac{\delta}{\delta A_{\bar{z}}^a} - \frac{k}{4\pi}A_z^a\big) \quad (7.40)$$

The holomorphic (or Bargmann) polarization condition is

$$\nabla\,\Phi = 0 \quad (7.41)$$

The solution of this condition are the wave functions of the form

$$\Phi = \exp\left(-\frac{k}{4\pi}\int A_{\bar{z}}^a A_z^a\right)\ \psi[A_{\bar{z}}^a] = e^{-\frac{1}{2}K}\ \psi[A_{\bar{z}}^a] \quad (7.42)$$

where K is the Kähler potential of (7.31). The states are represented by wave functionals $\psi[A_{\bar{z}}^a]$ which are holomorphic in $A_{\bar{z}}^a$. Further, the prequantum inner product can be retained as the inner product of the Hilbert space. Rewriting (7.38) using (7.42) we get the inner product as

$$\langle 1|2\rangle = \int [dA_{\bar{z}}^a\,dA_z^a]\,e^{-K(A_{\bar{z}}^a, A_z^a)}\,\psi_1^*\,\psi_2 \quad (7.43)$$

On the holomorphic wave functions,

$$A_z^a\,\psi[A_{\bar{z}}^a] = \frac{2\pi}{k}\frac{\delta}{\delta A_{\bar{z}}^a}\,\psi[A_{\bar{z}}^a] \quad (7.44)$$

As we have mentioned before, one has to make a reduction of the Hilbert space by imposing gauge invariance on the states, i.e., by setting the generator $F_{z\bar{z}}^a$ to zero on the wave functionals. This amounts to

$$\left[\left(D_{\bar{z}}\frac{\delta}{\delta A_{\bar{z}}}\right)^a - \frac{k}{2\pi}\partial_z A_{\bar{z}}^a\right]\psi[A_{\bar{z}}^a] = 0. \quad (7.45)$$

Consistent implementation of gauge invariance can lead to quantization requirements on the coupling constant k. For nonabelian groups G this is essentially the requirement that k should be an integer, based on the invariance of e^{iS} under homotopically nontrivial gauge transformations. It is also the same as the Dirac quantization condition (5.10). Further, once we impose the gauge invariance condition, the integration in (7.43) must be restricted to the gauge-invariant volume.

7.2 Argument for Quantization of k

We will now work out how the quantization of k arises, in some detail, staying within the geometric quantization framework. Since we are on S^2, the group of gauge transformations consists of maps from S^2 to G. We have chosen $G = SU(N)$, so obviously

$$\Pi_0(\mathfrak{G}_*) = \Pi_2(G) = 0, \qquad \Pi_1(\mathfrak{G}_*) = \Pi_3(G) = \mathbb{Z} \qquad (7.46)$$

The space of fields $\mathfrak{F}$ is an affine space with trivial topology. Therefore, the homotopy groups given above imply that

$$\Pi_1(\mathfrak{F}/\mathfrak{G}_*) = 0, \qquad \Pi_2(\mathfrak{F}/\mathfrak{G}_*) = \mathbb{Z} \qquad (7.47)$$

The nontriviality of $\Pi_2(\mathfrak{F}/\mathfrak{G}_*)$ arises from the nontrivial elements of $\Pi_1(\mathfrak{G}_*)$. Therefore consider a noncontractible loop C of gauge transformations,

$$C = g(x, \alpha), \qquad 0 \le \alpha \le 1, \quad \text{with} \quad g(x, 0) = g(x, 1) = 1 \qquad (7.48)$$

With the boundary condition given, $g(x, \alpha)$ may be considered as a map from S^3 to G. Such elements fall into homotopy classes corresponding to $\Pi_3(G) = \mathbb{Z}$. We can now use this $g(x, \alpha)$ to construct an example of a noncontractible two-surface in the gauge -invariant space $\mathfrak{F}/\mathfrak{G}_*$. We start with a square in the space of gauge potentials parmetrized by $0 \le \alpha, \sigma \le 1$ with the potentials given by

$$A(x, \alpha, \sigma) = (g \, A \, g^{-1} - \mathrm{d} g \, g^{-1}) \sigma + (1 - \sigma) A \qquad (7.49)$$

For our purpose, we can simplify this even further by taking $A = 0$, so that

$$A(x, \alpha, \sigma) = -\sigma \, \mathrm{d} g \, g^{-1} \qquad (7.50)$$

This potential goes to zero on the boundaries $\alpha = 0$ and $\alpha = 1$ and also on $\sigma = 0$. A goes to the pure gauge $-\mathrm{d} g g^{-1}$ at $\sigma = 1$, which is gauge-equivalent to $A = 0$. Thus the boundary corresponds to a single point on the quotient $\mathfrak{F}/\mathfrak{G}_*$ and we have a closed two-surface. This surface is noncontractible if we take $g(x, \alpha)$ to be a nontrivial element of $\Pi_3(G) = \mathbb{Z}$ since the contraction of the two-surface would constitute a homotopy mapping g to the identity; this is impossible if g belongs to a nontrivial element of $\Pi_3(G)$. Using this set of configurations in Ω and carrying out the integration over σ we get

$$\int \Omega = -2\pi \, k \, Q[g] \qquad (7.51)$$

where

$$Q[g] = -\frac{1}{24\pi^2} \int \text{Tr}(dgg^{-1})^3 \tag{7.52}$$

This quantity $Q[g]$ is the winding number (which is an integer) characterizing the class in $\Pi_1(\mathfrak{G}_*) = \Pi_3(G)$ to which g belongs. From (5.10) we know that the integral of Ω over any closed noncontractible two-surface in the phase space must be an integer. Thus we see that (7.51) and (7.52) lead to the requirement that the level number k of the Chern-Simons theory has to be an integer. (Even though we presented the arguments for quantization of the coefficient of the action for $\Sigma = S^2$, similar arguments and results hold more generally.)

7.3   The Ground State Wave Function

We now turn to the solution of (7.45). For this we introduce the Wess-Zumino-Witten (WZW) action given by [17, 25]

$$S_{\text{wzw}} = \frac{1}{8\pi} \int_\Sigma d^2x \sqrt{g}\, g^{ab} \text{Tr}(\partial_a K \partial_b K^{-1}) + \Gamma_{\text{wz}}[K]$$

$$\Gamma_{\text{wz}}[K] = \frac{i}{12\pi} \int_{\mathcal{M}^3} \text{Tr}(K^{-1}dK)^3 \tag{7.53}$$

The fields are matrices K which can generally belong to $GL(N, \mathbb{C})$. Also Σ is the two-dimensional space on which the fields are defined. Since it can in general be a curved manifold, we use the two-dimensional metric tensor g_{ab}. (g^{ab} is the inverse metric and g denotes the determinant of g_{ab} as a matrix.) (This model can be defined and used for fields on $\mathbb{R}^2$ as well, by choosing the boundary condition $K \to 1$ (or some fixed value independent of directions) as $|x| \to \infty$; topologically, such fields are equivalent to fields on the closed manifold S^2.)

The second term in the action, $\Gamma_{\text{wz}}[K]$, is the so-called Wess-Zumino term. It is defined in terms of integration over a three-dimensional space $\mathcal{M}^3$ which has Σ as its boundary. The integrand does not require metrical factors for the integration since it is a differential three-form. However, it requires an extension of the field K to the three-space $\mathcal{M}^3$. There can be many spaces $\mathcal{M}^3$ with the same boundary Σ, or equivalently, there can be many different ways to extend the fields to the three-space $\mathcal{M}^3$. The physical results of the theory are independent of how this extension is chosen, if we consider actions of the form $k\, S_{\text{wzw}}$ where k is an integer. By direct calculation, we can verify the Polyakov-Wiegmann identity [26][1]

---

[1] In accordance with the convention for A_z, $A_{\bar{z}}$ given above (7.30), here we use $\partial_z = \frac{1}{2}(\partial_1 + i\partial_2)$, $\partial_{\bar{z}} = \frac{1}{2}(\partial_1 - i\partial_2)$ in terms of real components.

$$\mathcal{S}_{\text{wzw}}[K\,h] = \mathcal{S}_{\text{wzw}}[K] + \mathcal{S}_{\text{wzw}}[h] - \frac{1}{\pi}\int_{\Sigma} \text{Tr}(K^{-1}\partial_{\bar{z}}K\,\partial_z h\,h^{-1}) \tag{7.54}$$

where we have used local complex coordinates. Now, in two dimensions, we can parametrize a nonabelian gauge potential as

$$A_z = -\partial_z M\,M^{-1}, \qquad A_{\bar{z}} = M^{\dagger-1}\partial_{\bar{z}}M^{\dagger} \tag{7.55}$$

where M is a complex matrix which may be taken to be in $SL(N,\mathbb{C})$ for gauge fields corresponding to the gauge group $SU(N)$. The identity (7.54) shows that

$$\delta\mathcal{S}_{\text{wzw}}(M^{\dagger}) = \mathcal{S}_{\text{wzw}}[M^{\dagger}\,(1+\theta)] - \mathcal{S}_{\text{wzw}}[M^{\dagger}] = \frac{1}{\pi}\int \text{Tr}(\partial_z A_{\bar{z}}\,\theta) \tag{7.56}$$

With $D_{\bar{z}}$ denoting the covariant derivative with respect to $A_{\bar{z}}$, we also have the identity

$$\partial_z A_{\bar{z}} = D_{\bar{z}}(M^{\dagger-1}\partial_z M^{\dagger}) \tag{7.57}$$

Notice that, since $\delta M^{\dagger} = M^{\dagger}\,\theta$, we may write $\theta = M^{\dagger-1}\delta M^{\dagger}$; further, from (7.55), $\delta A_{\bar{z}} = D_{\bar{z}}(M^{\dagger-1}\delta M^{\dagger}) = D_{\bar{z}}\theta$. Combining these relations with (7.57), we can simplify (7.56) as

$$D_{\bar{z}}\frac{\delta\mathcal{S}_{\text{wzw}}}{\delta A_{\bar{z}}^a} = \frac{1}{2\pi}\partial_z A_{\bar{z}}^a \tag{7.58}$$

where we have also evaluated the trace in terms of the components. Comparing this with (7.45), we see that we can solve it as

$$\psi(A_{\bar{z}}) = \mathcal{N}\,\exp\left(k\,\mathcal{S}_{\text{wzw}}[M^{\dagger}]\right) \tag{7.59}$$

The normalization factor $\mathcal{N}$ is to be fixed by using the inner product (7.43). There is only one state for this theory. On S^2, there are no degrees of freedom left for the Chern-Simons theory after one reduces to the physical configuration space. Thus there is only the vacuum state of the theory. What we have found is the expression for the ground state wave function in terms of the variables on $\mathfrak{F}$. If we consider higher genus Riemann surfaces, or two-manifolds with a boundary, then the Chern-Simons theory will have nontrivial degrees of freedom.

7.4 Abelian Theory on the Torus

We will now consider an Abelian Chern-Simons theory, with $G = U(1)$ and with Σ being a torus $S^1 \times S^1$. This will illustrate some of the topological features we mentioned. The torus can be described by $z = \xi_1 + \tau\xi_2$, where ξ_1, ξ_2 are real and have periodicity of $\xi_i \rightarrow \xi_i +$ integer, and τ, which is a complex number, is the modular parameter of the torus. The metric on the torus is $ds^2 = |d\xi_1 + \tau d\xi_2|^2$. The two basic noncontractible cycles of the torus are usually labelled as the α and β cycles. Further the torus has a holomorphic one-form ω with

$$\int_\alpha \omega = 1, \qquad \int_\beta \omega = \tau \tag{7.60}$$

Since ω is a zero mode of $\partial_{\bar{z}}$, we can parametrize $A_{\bar{z}}$ as[2]

$$A_{\bar{z}} = \partial_{\bar{z}}\chi + i\frac{\pi \bar{\omega}_{\bar{z}}}{\mathrm{Im}\tau} a \tag{7.61}$$

where χ is a complex function and a is a complex number corresponding to the value of $A_{\bar{z}}$ along the zero mode of ∂_z. Also $\mathrm{Im}\tau$ denotes the imaginary part of τ.

For this space $\Pi_0(\mathfrak{G}_*) = \mathbb{Z} \times \mathbb{Z}$, because the gauge transformations $g_{m,n}$ can have nontrivial winding numbers m, n around the two cycles. Consider one connected component of $\mathfrak{G}_*$, say $\mathfrak{G}_{m,n}$. A homotopically nontrivial $U(1)$ transformation can be written as $g_{m,n} = e^{i\alpha}\, e^{i\theta_{m,n}}$, where $\alpha(z, \bar{z})$ is a homotopically trivial gauge transformation and

$$\theta_{m,n} = \frac{i\pi}{\mathrm{Im}\tau}\left[m\int^z \bar{\omega} - \omega + n\int^z \tau\bar{\omega} - \bar{\tau}\omega \right], \qquad m, n \in \mathbb{Z} \tag{7.62}$$

With the parametrization of $A_{\bar{z}}$ as in (7.61), the effect of this gauge transformation can be represented as

$$\chi \to \chi + \alpha, \qquad a \to a + m + n\tau \tag{7.63}$$

The real part of χ can be set to zero by an appropriate choice of α. (The imaginary part also vanishes when we impose the condition $F_{z\bar{z}} = 0$.) The physical subspace (which has only the zero modes left after reduction) is given by the values of a modulo the transformation (7.63), or in other words,

$$\text{Physical space for zero modes} \equiv \mathfrak{C} = \frac{\mathbb{C}}{\mathbb{Z} + \tau\mathbb{Z}} \tag{7.64}$$

This space is known as the Jacobian variety of the torus. It is also a torus and therefore we see that the phase space $\mathfrak{C}$ has nontrivial Π_1 and $\mathcal{H}^2$. In particular, $\Pi_1(\mathfrak{C}) = \mathbb{Z} \times \mathbb{Z}$ and this leads to two angular parameters φ_α and φ_β which are the phases the wave functions acquire under the gauge transformation $g_{1,1}$. The symplectic two-form can be written as

---

[2] In this section, ω will denote a one-form obeying (7.60), not the Kähler two-form as in previous sections.

$$\Omega = \frac{k}{4\pi} \int \bar{\partial}\delta\chi \wedge \partial\delta\bar{\chi} + \frac{k\pi}{4} \frac{d\bar{a} \wedge da}{\mathrm{Im}\tau} \int_{\Sigma} \frac{\bar{\omega} \wedge \omega}{\mathrm{Im}\tau}$$

$$= \Omega_{\chi} - \mathrm{i}\frac{k\pi}{2} \frac{d\bar{a} \wedge da}{\mathrm{Im}\tau} \tag{7.65}$$

(Here we are using the notation of holomorphic and antiholomorphic exterior derivatives, so that $\bar{\partial}\delta\chi = \partial_{\bar{z}}\delta\chi\, d\bar{z}$, etc.) Integrating the zero mode part over the physical space of zero modes $\mathfrak{C}$, we get

$$\int_{\mathfrak{C}} \Omega = k\,\pi \tag{7.66}$$

showing that k must be quantized as an even integer for $U(1)$ fields on the torus due to (5.10).[3]

The modular parameter of the torus is subject to the so-called modular transformations which are homotopically nontrivial diffeomorphisms of the torus. The vacuum angles change under such transformations and can eventually be set to zero. To continue with the quantization, we focus on the zero modes for which the symplectic potential can be written as

$$\mathcal{A} = -\frac{\pi k}{4} \frac{(\bar{a} - a)(\tau\, d\bar{a} - \bar{\tau}\, da)}{(\mathrm{Im}\tau)^2} \tag{7.67}$$

The polarization condition then becomes

$$\left[\frac{\partial}{\partial\bar{a}} + \mathrm{i}\frac{\pi k}{4} \frac{(\bar{a} - a)\tau}{(\mathrm{Im}\tau)^2} \right] \Psi = 0 \tag{7.68}$$

with the solution

$$\Psi = \exp\left[-\mathrm{i}\frac{\pi k}{8} \frac{(\bar{a} - a)^2\tau}{(\mathrm{Im}\tau)^2} \right] f(a) \tag{7.69}$$

where $f(a)$ is holomorphic in a. Under the gauge transformation (7.63) we find

$$\Psi(a + m + n\tau) = \exp\left[-\mathrm{i}\frac{\pi k(\bar{a} - a)^2\tau}{8(\mathrm{Im}\tau)^2} - \frac{\pi kn(\bar{a} - a)\tau}{2\mathrm{Im}\tau} + \mathrm{i}\frac{\pi k\tau n^2}{2} \right] f(a + m + n\tau) \tag{7.70}$$

---

[3] Since there have been different statements on this point in the literature, a comment might be in order. In geometric quantization, we are considering the wave functions as sections of a line bundle. This means that each quantum state has a wave function which is a complex number. One can avoid the quantization condition on k for the Abelian theory if one is willing to go beyond this and allow for more general or multicomponent wave functions (for each state). However, the interpretation of the theory in such a situation will be very different [27].

Under this gauge transformation $\mathcal{A}$ changes by $d\Lambda_{m,n}$ where

$$\Lambda_{m,n} = i\frac{\pi k n (\tau \bar{a} - \bar{\tau}a)}{2 \operatorname{Im}\tau} \qquad (7.71)$$

The change in Ψ should thus be given by $\exp(i\Lambda_{m,n})\Psi$; requiring the transformation (7.70) to be equal to this, we get

$$f(a + m + n\tau) = \exp\left[-i\frac{\pi k n^2 \tau}{2} - i\pi k n a\right] f(a) \qquad (7.72)$$

This transformation rule shows that $f(a)$ is a Jacobi Θ-function. The operator $\bar{a}$ is realized on these functions $f(a)$ as

$$\bar{a} f(a) = \left[\frac{2 \operatorname{Im}\tau}{k\pi}\frac{\partial}{\partial a} + a\right] f(a) \qquad (7.73)$$

The inner product for the wave functions of the zero modes is

$$\langle f|g\rangle = \int \exp\left[-\frac{\pi k \bar{a}a}{2 \operatorname{Im}\tau} + \frac{\pi k \bar{a}^2}{4 \operatorname{Im}\tau} + \frac{\pi k a^2}{4 \operatorname{Im}\tau}\right] \bar{f}g \qquad (7.74)$$

It is then convenient to absorb the holomorphic part of the exponent into the wave function defining the new set of holomorphic wave functions

$$\Phi \equiv \exp\left[\frac{\pi k a^2}{4 \operatorname{Im}\tau}\right] f(a) = \exp\left[\frac{\pi k a^2}{4 \operatorname{Im}\tau}\right] \Theta(a) \qquad (7.75)$$

On these functions, $\bar{a}$ acts as

$$\bar{a} = \frac{2 \operatorname{Im}\tau}{k\pi}\frac{\partial}{\partial a} \qquad (7.76)$$

The key point we wanted to illustrate here is the use of the homotopically nontrivial gauge transformations.

Problems

7.1 Calculate $\Omega(A^g) - \Omega(A)$ for finite transformations, i.e., obtain (7.37).

7.2 Derive the Polyakov-Wiegmann identity given in (7.54).

7.3 The WZW action can be quantized as a 1+1 dimensional field theory in its own right. In lightcone coordinates $u = (t - x)/\sqrt{2}$, $v = (t + x)/\sqrt{2}$, the action is

$$\mathcal{S} = -\frac{k}{4\pi}\int \operatorname{Tr}(\partial_u g g^{-1} \partial_v g g^{-1}) + \Gamma_{\mathrm{wz}}$$

Identify the canonical two-form. Show that left translations of g, i.e., $g \to (1 + (-it_a\theta^a))g$ are generated by $J_v^a = (k/4\pi)(\partial_v g g^{-1})^a$. Obtain also the commutation rules

$$[J_v(\theta), J_v(\varphi)] = i J_v(\theta \times \varphi) - i\frac{k}{4\pi}\int \partial_v \theta^a \varphi^a.$$

Open Access This chapter is licensed under the terms of the Creative Commons Attribution 4.0 International License (http://creativecommons.org/licenses/by/4.0/), which permits use, sharing, adaptation, distribution and reproduction in any medium or format, as long as you give appropriate credit to the original author(s) and the source, provide a link to the Creative Commons license and indicate if changes were made.

The images or other third party material in this chapter are included in the chapter's Creative Commons license, unless indicated otherwise in a credit line to the material. If material is not included in the chapter's Creative Commons license and your intended use is not permitted by statutory regulation or exceeds the permitted use, you will need to obtain permission directly from the copyright holder.

Chapter 8
θ-Vacua in a Nonabelian Gauge Theory

Consider a nonabelian gauge theory in four spacetime dimensions, the gauge group is some compact Lie group G. For specificity we may consider the Yang-Mills theory defined by the action

$$S = -\frac{1}{4e^2} \int d^4x \, F^{a\mu\nu} F^a_{\mu\nu}$$
$$F^a_{\mu\nu} = \partial_\mu A^a_\nu - \partial_\nu A^a_\mu + f^{abc} A^b_\mu A^c_\nu \tag{8.1}$$

Here e is the coupling constant and f^{abc} are the structure constants defined by $[t^b, t^c] = i f^{abc} t^a$, where $\{t^a\}$ are the generators of the Lie algebra of G. We can choose the gauge where $A_0 = 0$ so that there are only the three spatial components of the gauge potential, namely, A_i, considered as an antihermitian Lie algebra-valued vector field. The choice $A_0 = 0$ does not completely fix the gauge, one can still do gauge transformations which are independent of time. These are given by

$$A_i \rightarrow A'_i = g A_i g^{-1} - \partial_i g \, g^{-1} \tag{8.2}$$

The Yang-Mills action gives the symplectic two-form as

$$\Omega = \int d^3x \, \delta E^a_i \, \delta A^a_i = -2 \int d^3x \, \text{Tr}\,(\delta E_i \, \delta A_i) \tag{8.3}$$

where E^a_i is the electric field $\partial_0 A^a_i / e^2$, along the Lie algebra direction labeled by a. The gauge transformation of E_i is $E_i \rightarrow g E_i g^{-1}$. By combining this with the transformation (8.2), we identify the vector field generating infinitesimal gauge transformations, with $g \approx 1 + \varphi$, as

$$\xi = - \int d^3x \left[(D_i \varphi)^a \frac{\delta}{\delta A^a_i} + [E_i, \varphi]^a \frac{\delta}{\delta E^a_i} \right] \tag{8.4}$$

© The Author(s) 2024
V. P. Nair, *Geometric Quantization and Applications to Fields and Fluids*,
SpringerBriefs in Physics, https://doi.org/10.1007/978-3-031-65801-3_8

This leads to

$$i_\xi \Omega = -\delta \int \mathrm{d}^3 x \ \left[-(D_i \varphi)^a E_i^a \right] \tag{8.5}$$

The generator of time-independent gauge transformations is thus

$$G(\varphi) = -\int \mathrm{d}^3 x \ (D_i \varphi)^a E_i^a \tag{8.6}$$

For transformations which go to the identity at spatial infinity, $G^a = (D_i E_i)^a$. This is Gauss law, one of the Yang-Mills equations of motion. As before, it is to be viewed as a condition on the allowed initial data and enforces a reduction of the phase space to gauge-invariant variables. We again define the space of fields and gauge transformations as

$$\mathfrak{F} = \left\{ \text{Space of gauge potentials } A_i \right\} \tag{8.7}$$

$$\mathfrak{G}_* = \left\{ \begin{array}{c} \text{Space of gauge transformations } g(\boldsymbol{x}) : \mathbb{R}^3 \to G \\ \text{such that } g \to 1 \text{ as } |\boldsymbol{x}| \to \infty \end{array} \right\} \tag{8.8}$$

The transformations $g(\boldsymbol{x})$ which go to a constant element g_∞ which is not necessarily equal to 1 act as a Noether symmetry. The states fall into unitary irreducible representations of such transformations, which are isomorphic to the gauge group G, upto $\mathfrak{G}_*$-transformations. The true gauge freedom is only $\mathfrak{G}_*$. The physical configuration space of the theory is thus $\mathfrak{C} = \mathfrak{F}/\mathfrak{G}_*$.[1]

With the boundary condition on the g's, the gauge functions are equivalent to a map from S^3 to G, and hence there are homotopically distinct transformations corresponding to the fact that $\Pi_3(G) = \mathbb{Z}$. (In other words, $\Pi_0(\mathfrak{G}_*) = \mathbb{Z}$.) These can be labeled by the winding number $Q[g]$ given in (7.52). We can write $\mathfrak{G}_*$ as the sum of different components, each of which is connected and is characterized by the winding number Q, i.e.,

$$\mathfrak{G}_* = \sum_{Q=-\infty}^{+\infty} \oplus \ \mathfrak{G}_Q \tag{8.9}$$

where each $\mathfrak{G}_Q$ consists of all maps with winding number Q. $\mathfrak{G}_Q$ and $\mathfrak{G}_{Q'}$ are disconnected from each other for $Q \neq Q'$, since if they are connected, $g_Q \in \mathfrak{G}_Q$ and $g_{Q'} \in \mathfrak{G}_{Q'}$ should be homotopically deformable to each other and this is impossible since $Q \neq Q'$. One can easily check that $Q[g \, g'] = Q[g] + Q[g']$ and hence this structure is isomorphic to the additive group of integers $\mathbb{Z}$. The space of gauge potentials $\mathfrak{F}$ is an affine space and is topologically trivial. Combining these facts, we see that the configuration space has noncontractible loops, with $\Pi_1(\mathfrak{C}) = \Pi_3(G) = \mathbb{Z}$.

---

[1] For further elaborations on this question, see [6, 20] and references therein.

An example of a noncontractible loop in $\mathfrak{C}$ is as follows. Let $g_1(x)$ be a gauge transformation with winding number equal to 1 and consider the line in $\mathfrak{F}$ given by

$$A_i(x, \tau) = A_i(x)(1 - \tau) + A_i^{g_1} \tau \tag{8.10}$$

for $0 \le \tau \le 1$, or more generally

$$A_i(x, \tau) \quad \text{with} \quad A_i(x, 0) = A_i(x), \quad A_i(x, 1) = A_i^{g_1}(x) \tag{8.11}$$

where $A_i^{g_1}$ is the gauge transform of A_i by $g_1(x)$. This is an open path in $\mathfrak{F}$. But since $A_i^{g_1}$ is the gauge transform of A_i, both configurations A_i and $A_i^{g_1}$ represent the same point in $\mathfrak{C} = \mathfrak{F}/\mathfrak{G}_*$. Thus $A_i(x, \tau)$ describes a closed loop in $\mathfrak{C}$. If this loop is contractible, we can deform the trajectory to a curve purely along the gauge flow directions which connects $g = 1$ to $g_1(x)$. This would imply that $g_1(x)$ is smoothly deformable to the identity. But this is impossible from our discussion of the structure of $\mathfrak{G}_*$. In turn this implies that $A_i(x, \tau)$ of (8.11) is a noncontractible loop. By considering other values of the winding number, we can easily establish that $\Pi_1(\mathfrak{C}) = \mathbb{Z}$. Our general discussion from Chap. 5 shows that there must be a vacuum angle θ which appears in the quantum theory. We can now see how this emerges by writing out the symplectic potential [28, 29].

We will first construct a flat potential on the space of fields. For this we start with the instanton number which is given, for a four-dimensional potential, by

$$
\begin{aligned}
\nu[A] &= -\frac{1}{32\pi^2} \int d^4x \, \mathrm{Tr}\left(F_{\mu\nu} F_{\alpha\beta}\right) \epsilon^{\mu\nu\alpha\beta} \\
&= \frac{1}{16\pi^2} \int d^4x \, \epsilon^{ijk} E_i^a F_{jk}^a
\end{aligned}
\tag{8.12}
$$

The density in the above integral is a total derivative in terms of the potential A, but it cannot be written as a total derivative in terms of gauge-invariant quantities. $\nu[A]$ is an integer for any field configuration which is nonsingular up to gauge transformations. It is possible to construct configurations which have a nonzero value of ν and which are nonsingular; these are instantons in a general sense.[2] An example of a $\nu = -1$ configuration, for $G = SU(2)$, is

$$A_\mu(x) = \frac{x^2}{x^2 + \alpha^2} \, \omega^{-1} \partial_\mu \omega, \qquad \omega = \frac{x_4 + i\,\sigma \cdot x}{\sqrt{x^2}} \tag{8.13}$$

---

[2] There is a more specific sense in which the word instanton is used; it applies to self-dual solutions of the Yang-Mills equations which further have $\nu[A] \neq 0$.

For our purpose, we can transform this to the gauge with $A_0 = A_4 = 0$ obtaining

$$
A_i = U \left(\frac{x^2}{x^2 + \alpha^2} \omega^{-1} \partial_i \omega \right) U^{-1} - \partial_i U \, U^{-1}, \qquad\qquad U = \exp \left(i \boldsymbol{\sigma} \cdot \hat{x} \, \rho \right)
$$

$$
\rho = \frac{|\boldsymbol{x}|}{\sqrt{|\boldsymbol{x}|^2 + \alpha^2}} \left[\arctan \left(\frac{x_4}{\sqrt{|\boldsymbol{x}|^2 + \alpha^2}} \right) + \frac{\pi}{2} \right], \qquad \frac{\partial \rho}{\partial x_4} = \frac{|\boldsymbol{x}|}{x^2 + \alpha^2} \qquad (8.14)
$$

Again, in these equations, σ_i are the Pauli matrices and the path is parametrized by x_4, $-\infty \leq x_4 \leq \infty$. $x^2 = \boldsymbol{x}^2 + x_4^2$. Since x_4 parametrizes the path, we see that $\nu[A]$ can be written as

$$
\nu[A] = \oint K[A]
$$

$$
K[A] = -\frac{1}{8\pi^2} \int d^3x \, \epsilon^{ijk} \mathrm{Tr}(\delta A_i F_{jk}) = \frac{1}{16\pi^2} \int d^3x \, \epsilon^{ijk} \delta A_i^a F_{jk}^a \qquad (8.15)
$$

The integral of K, which is a one-form on the configuration space, around a closed curve is the instanton number ν and is nonzero, in particular, for the loop corresponding to (8.14) for which $\nu = -1$. We can also see that the one-form $K[A]$ on $\mathfrak{C}$ is closed in the following way.

$$
\begin{aligned}
\delta K[A] &= -\frac{1}{8\pi^2} \int d^3x \; \delta \Big(\mathrm{Tr} \left(F_{jk} \delta A_i \right) \Big) \epsilon^{ijk} \\
&= -\frac{1}{4\pi^2} \int d^3x \; \mathrm{Tr} \left((D_j \delta A_k) \, \delta A_i \right) \epsilon^{ijk} \\
&= -\frac{1}{4\pi^2} \int d^3x \; \mathrm{Tr} \left(\partial_j \delta A_k \, \delta A_i + [A_j, \delta A_k] \delta A_i \right) \epsilon^{ijk} \\
&= 0
\end{aligned} \qquad (8.16)
$$

In the last step we have used the antisymmetry of the expression under permutation of δ's, cyclicity of the trace and have carried out an integration by parts. We see from the above discussion that $K[A]$ is a closed one-form, but it is not exact since its integral around the closed curves can be nonzero.

With this flat potential on $\mathfrak{C}$, we can construct a general solution for the symplectic potential corresponding to the Ω in (8.3) as

$$
\mathcal{A} = \int d^3x \; E_i^a \delta A_i^a + \theta \, K[A] \qquad (8.17)
$$

Use of this potential will lead to a quantum theory where we need the parameter θ, in addition to other parameters such as the coupling constant, to characterize the theory. The potential $\mathcal{A}$ in (8.17) is obtained from an action

$$\mathcal{S} = -\frac{1}{4e^2} \int d^4x \ F^a_{\mu\nu} F^{a\mu\nu} + \theta\nu[A] \tag{8.18}$$

This shows that the effect of using (8.17) can be reproduced in the functional integral approach by using the action (8.18). Since it is $\exp(i\mathcal{S})$ which is important, we see that θ is an angle with values $0 \le \theta < 2\pi$. Alternatively, we can see that one can formally eliminate the θ-term in $\mathcal{A}$ by making a transformation $\Psi \to \exp(i\theta\Lambda)\Psi$ where

$$\Lambda = -\frac{1}{8\pi^2} \int \mathrm{Tr}\left(A\,dA + \frac{2}{3}A\,A\,A\right) \tag{8.19}$$

Notice that $2\pi\Lambda$ is the Chern-Simons action (7.1) for $k = 1$. Λ is not invariant under homotopically nontrivial transformations. The wave functions get a phase equal to $e^{i\theta Q}$ under a transformation with winding number equal to Q, showing that θ can be restricted to the interval indicated above. This is in agreement with our discussion after Eq. (5.4).

Problem

8.1 Calculate $\nu[A]$ for the instanton in (8.13) and (8.14).

Open Access This chapter is licensed under the terms of the Creative Commons Attribution 4.0 International License (http://creativecommons.org/licenses/by/4.0/), which permits use, sharing, adaptation, distribution and reproduction in any medium or format, as long as you give appropriate credit to the original author(s) and the source, provide a link to the Creative Commons license and indicate if changes were made.

The images or other third party material in this chapter are included in the chapter's Creative Commons license, unless indicated otherwise in a credit line to the material. If material is not included in the chapter's Creative Commons license and your intended use is not permitted by statutory regulation or exceeds the permitted use, you will need to obtain permission directly from the copyright holder.

Chapter 9
Fractional Statistics in Quantum Hall Effect

The quantum Hall effect is an important physical phenomenon which has been analyzed both theoretically and experimentally for a few decades by now [30]. In this chapter, we will discuss this effect in relation to geometric quantization. We will also see how fractional statistics for quasiparticles is related to nontrivial $\mathcal{H}^1(M, \mathbb{R})$ of the phase space.

9.1 Quantum Hall Effect and the Landau Problem

When an electric field is applied, the electrons in a conducting material will move in response, leading to an electric current in the direction of the applied field. If there is also an applied magnetic field (which is usually taken to be uniform), the Lorentz force due to this will deflect electrons in a direction transverse to their velocity (i.e., transverse to the applied electric field) and this can create a voltage and a current in the transverse direction. This is Hall effect. We can write the transverse current J_i, also referred to as the Hall current, in the form

$$J_i = \sigma_{ij} E_j \tag{9.1}$$

where E_j is the applied field. The indices i, j take values 1, 2, corresponding to a two-dimensional plane, the magnetic field is taken to be along the third direction. σ_{ij} is the Hall conductivity and, classically, it is proportional to the magnetic field. At very low temperatures, however, a plot of σ_{ij} versus the magnetic field B shows a series of plateaux, where the value of σ_{ij} is independent of B for a certain range of B, with the system making a transition to another plateau as B is increased or decreased beyond this range. The general expression is

© The Author(s) 2024
V. P. Nair, *Geometric Quantization and Applications to Fields and Fluids*,
SpringerBriefs in Physics, https://doi.org/10.1007/978-3-031-65801-3_9

$$J_i = -\nu \frac{e^2}{2\pi} \epsilon_{ij} E_j \tag{9.2}$$

where e is the charge of the electron. The coefficient ν in (9.2) can take integer values (1, 2, 3, etc.) or certain rational fractional values. These cases are referred to as the integer and fractional quantum Hall effects (QHE), respectively. Quantum effects are crucial for the plateaux behavior with quantized values of ν, hence the qualification as quantum Hall effect.

The quantization of ν is seen to be very robust, insensitive to impurities in the sample (for some range of densities for the impurities) and can be related to the integrals of certain Chern classes associated with the band structure of electrons in the material.

The dynamics of charged particles in a uniform magnetic field is known as the Landau problem. In the specific context of QHE, the electrons are the ones in the conduction band, i.e., the freely movable electrons in the material. The general expectation is that the analysis in terms of the Landau problem neglecting the mutual Coulomb interaction of the electrons suffices for the integer QHE. The Coulomb interaction is expected to be significant for obtaining the fractional QHE states.

Our aim here will be to highlight certain features of the QHE which overlap with considerations of geometric quantization. Towards this we start by considering the Landau problem of charged particles moving on a two-sphere $S^2 = SU(2)/U(1)$ [12, 13]. As before, we can parametrize the sphere by $g \in SU(2)$ with the identification $g \sim gh$, $h \in U(1) \subset SU(2)$. Right translation operators $R_\pm$ on g, defined by

$$R_\pm g = g t_\pm, \qquad t_\pm = \tfrac{1}{2}(\sigma_1 \pm i\sigma_2), \tag{9.3}$$

are the translation operators on the sphere. The covariant derivatives are given by $D_\pm = i R_\pm/r$, where r is the radius of the sphere. The Hamiltonian for the charged particle is proportional to the covariant Laplacian and has the form

$$H = -\frac{1}{2m} D^2 = -\frac{1}{4m}(D_+ D_- + D_- D_+) = \frac{1}{4mr^2}(R_+ R_- + R_- R_+)$$
$$= \frac{1}{2mr^2}(R_+ R_- - R_3) \tag{9.4}$$

where m is the mass of the particle.

In viewing S^2 as $SU(2)/U(1)$, the $U(1)$ subgroup (generated by R_3) is the local isotropy group of frame rotations, while $SU(2)$, which includes translations as well, gives the full isometry group. Thus the curvature of S^2 takes values in the Lie algebra of $U(1)$ and is a constant in the tangent frame basis for S^2. A magnetic field which is uniform will be proportional to the curvature. Since the commutator of covariant derivatives is proportional to the curvatures, the relation $[R_+, R_-] = 2R_3$ shows that the value of R_3 is proportional to the background magnetic field. (There can be an additional term proportional to the spin times the spatial curvature if the particle has spin.) Taking the eigenvalue of R_3 as $-\frac{1}{2}n$ for some integer n, we see that, for a spinless particle,[1]

$$-n = 2R_3 = [R_+, R_-] = -r^2[D_+, D_-] = -r^2(2eB) \tag{9.5}$$

This identifies the magnetic field as

$$B = \frac{n}{2er^2}, \quad e\left[\frac{\oint B \cdot dS}{4\pi}\right] = \frac{n}{2} \tag{9.6}$$

The condition on the integral of B is consistent with the Dirac quantization condition on the charge of a monopole. A uniform magnetic field corresponds to a monopole at the origin of $\mathbb{R}^3$ if we think of the sphere as embedded in $\mathbb{R}^3$. The whole discussion here is agreement with similar considerations in Chap. 6.

As for the eigenfunctions of H, consider functions on $SU(2)$; i.e., functions of $g = \exp(i\sigma_a \theta^a/2)$. By the Peter-Weyl theorem, they can be expanded as

$$\psi = \sum_{j,p,w} C_{p,w}^{(j)} \langle j, p|e^{i\hat{J}\cdot\theta}|j, w\rangle = \sum_{j,p,w} C_{p,w}^{(j)} \mathcal{D}_{pw}^{(j)}(g) \tag{9.7}$$

We can then reduce this set by the required conditions on the states. The requirement on R_3 implies that we should choose $|j, w\rangle = |j, -n/2\rangle$. The eigenfunctions are thus

$$\psi_p^{(j)} = \mathcal{N}\langle j, p|e^{i\hat{J}\cdot\theta}|j, -n/2\rangle \tag{9.8}$$

The action of H on this identifies the energy eigenvalues as

$$E_q = \frac{n}{2mr^2}(q + \tfrac{1}{2}) + \frac{q(q+1)}{2mr^2} \tag{9.9}$$

where we write $j = (n/2) + q$, $q = 0, 1, 2$, etc. Equations (9.8) and (9.9) give the solution of the Landau problem on the sphere.

Defining the left translation operator by $L_a\, g = t_a\, g$, we see that $[L_a, R_b] = 0$, so that we also have

$$[L_a, H] = 0 \tag{9.10}$$

Thus the left action of $SU(2)$ on the group element g is a symmetry of the Hamiltonian and leads to degeneracy of the energy levels. In fact, notice that each wave function in (9.8) transforms as the spin-j representation of $SU(2)$, corresponding to left translations of the group element; i.e.,

$$\psi_p^{(j)}(Ug(\theta)) = \langle j, p|U|j, r\rangle \psi_r^{(j)} \tag{9.11}$$

---

[1] We take $D_1 = \partial_1 + iA_1$, $D_2 = \partial_1 + iA_2$ to relate B to the conventional definitions.

These left translations are referred to as magnetic translations in the context of the Landau problem. Since we have a uniform magnetic field, there should be a symmetry under translations on the sphere. The left translations give the explicit realization of this symmetry.

Obviously, from the expression for the energy levels in (9.9), $q = 0$ corresponds to the lowest Landau level. From the form of the Hamiltonian in (9.4), we also see that the lowest Landau level should satisfy

$$R_- \psi = 0 \tag{9.12}$$

Thus the state $|j, w\rangle$ in (9.7), for the lowest level, not only has R_3 equal to $-\frac{1}{2}n$, it is also the lowest weight state of the representation, so we can make the identification $j = \frac{n}{2}$, i.e., $q = 0$. The lowest Landau level has a degeneracy equal to $2j + 1 = n + 1$. It is easy to see that the states corresponding to the lowest Landau level agree with the geometric quantization on the two-sphere carried out in Chap. 6. The wave functions (9.8), with the parametrization of g as in (6.44), become

$$\psi_p = \sqrt{\frac{(n+1)!}{p!(n-p)!}} \frac{z^p}{(1+\bar{z}z)^{\frac{n}{2}}} \tag{9.13}$$

These are normalized as

$$\int \frac{\omega}{2\pi} \psi_p^* \psi_{p'} = \delta_{pp'}, \qquad \omega = i\frac{dz \wedge d\bar{z}}{(1+\bar{z}z)^2} \tag{9.14}$$

(ω is the Kähler two-form on S^2.) We may restate this correspondence as follows.

1. The Hilbert space of states in the lowest Landau level can be obtained by geometric quantization of the symplectic form $\Omega = -i\frac{n}{2}\mathrm{Tr}(\sigma_3 g^{-1}dg\, g^{-1}dg)$ on $S^2 = SU(2)/U(1)$.
2. The two-sphere represents the coordinate space of the particle from the point of view of the Landau problem; it becomes the phase space for dynamics in the lowest Landau level.
3. The condition $R_- \psi = 0$ selects the lowest Landau level for the Hamiltonian (proportional to the Laplace operator); it becomes the polarization condition from the point of view of the geometric quantization on S^2.

The wave functions (9.8) are the single particle wave functions. Quantum Hall effect is a many-particle phenomenon. If the particles are fermions (as they are in the physical situation since they are electrons), the many-particle wave function must be antisymmetric under permutation of particle positions. If the lowest Landau level is fully occupied (corresponding to $\nu = 1$), there should be $N = n + 1$ particles and the many-particle wave function is given by

$$\Psi(x_1, x_2, \ldots, x_N) = \frac{1}{\sqrt{N!}} \begin{vmatrix} \psi_1(x_1) & \psi_2(x_1) & \cdots & \psi_N(x_1) \\ \psi_1(x_2) & \psi_2(x_2) & \cdots & \psi_N(x_2) \\ \vdots & & \cdots & \vdots \\ \psi_1(x_N) & \psi_2(x_N) & \cdots & \psi_N(x_N) \end{vmatrix} \qquad (9.15)$$

It is easy to see the structure of this function by using the form of ψ_p written in terms of the homogeneous coordinates u_α, $\alpha = 1, 2$. The single particle wave function (9.13) is

$$\psi = \mathcal{N} u_{\alpha_1} u_{\alpha_2} \cdots u_{\alpha_n} \qquad (9.16)$$

This has n factors of u_α, with $N = n + 1$ distinct functions for different choices of $\{\alpha_1 \alpha_2 \cdots \alpha_n\}$. Notice that for two particles labeled as i, j, the combination $\epsilon_{\alpha\beta} u_\alpha^{(i)} u_\beta^{(j)}$ is antisymmetric under $i \leftrightarrow j$ and is invariant under a common (left) translation $u_\alpha \to U_{\alpha\gamma} u_\gamma$. In the wave function (9.15), u_α for each particle must appear n times, since the single particle function (9.16) has n factors of u_α, and further we must have antisymmetry under exchange of labels. These two features show that (9.15) is of the form

$$\Psi(x_1, x_2, \ldots, x_N) = \mathcal{N}' \prod_{i<j} [\epsilon_{\alpha\beta} u_\alpha^{(i)} u_\beta^{(j)}]$$

$$= \mathcal{N}' \prod_k \frac{1}{(1 + \bar{z}_k z_k)^{\frac{n}{2}}} \prod_{i<j} (z_i - z_j) \qquad (9.17)$$

This is known as the Laughlin wave function for the $\nu = 1$ QHE state on the sphere. Essentially all physical properties of the $\nu = 1$ state can be derived from this wave function.

It is also useful to consider the large radius limit of the sphere and obtain results for the two-dimensional plane. This can be done by the scaling

$$z, \bar{z} \to \frac{1}{2r} z, \ \frac{1}{2r} \bar{z}, \qquad r^2 = \frac{n}{2eB} \qquad (9.18)$$

where $z, \bar{z}$ on the right hand side are the coordinates on the plane. Taking $r^2 \to \infty$ at fixed eB, we find

$$\Psi(x_1, x_2, \ldots, x_N) = \mathcal{N}'' e^{-\frac{eB}{4} \sum \bar{z}_k z_k} \prod_{i<j} (z_i - z_j) \qquad (9.19)$$

This is the Laughlin wave function for $\nu = 1$ on the plane. In comparing this with the coherent states discussed in Chap. 6, we see that the exponential factor is just $\exp(-\frac{1}{2}K)$, and we can identify the parameter κ as $2/(eB)$. The large B limit is thus the analogue of the semiclassical limit of small κ.

9.2   Excitations in Fractional QHE

As mentioned earlier, there are also QHE states corresponding to fractional values of ν in (9.2). There are several such possibilities including the so-called Laughlin states, the Jain states, etc. [30]. We will focus here on Laughlin states with $\nu = \frac{1}{2p+1}$, $p = 1, 2, 3$, etc. A wave function which gives an excellent description of the physics of such states is

$$\Psi_{\text{Laughlin}} = \mathcal{N} \exp\left(-\frac{1}{2}\sum_{i=1}^{N} \bar{z}_i z_i\right) \prod_{1\leq i<j\leq N} (z_i - z_j)^{2p+1} \qquad (9.20)$$

where $z = x_1 + ix_2$, and the subscript, as before, refers to the particle. This is the wave function on the plane. While keeping factors of eB was important in going from the sphere to the plane, we now rescale the coordinates as $z \to \sqrt{2/eB}\, z$ to eliminate explicit factors of eB. (It should be noted that, so far, this wave function has not been derived from fundamental principles. Using (9.20) as an ansatz, one can show numerically that it is a good approximation to an eigenstate of the many-particle Hamiltonian including the Coulomb interaction of the electrons.) The wave function (9.20) leads to an electric current of the form

$$\langle J_i \rangle = -\nu \frac{e^2}{2\pi} \epsilon_{ij} E_j, \qquad \nu = \frac{1}{2p+1} \qquad (9.21)$$

This corresponds to the observed Hall conductivity, quantized as the reciprocals of odd integers.

In Chap. 5, we have discussed how the nontrivial connectivity of the phase space can lead to physical consequences via vacuum angles or via flat connections (which can modify the symplectic one-form but not the two-form). We will now show that excitations in the fractional quantum Hall effect will provide an example of how the nontrivial connectivity can affect the physics.

Among the excited states of the system are hole-like excitations, sometimes referred to as quasiparticles, with a wave function of the form

$$\Psi_{\text{hole}} = \prod_{i=1}^{N} (z_i - w)\, \Psi_{\text{Laughlin}}$$

$$= \mathcal{N} \prod_{i=1}^{N} (z_i - w)\, \exp\left(-\frac{1}{2}\sum_{i=1}^{N} \bar{z}_i z_i\right) \prod_{1\leq i<j\leq N} (z_i - z_j)^{2p+1} \qquad (9.22)$$

where w is the position of the hole. We want to briefly consider the statistics of such hole-like excitations in fractional quantum Hall effect. We can do this in an effective description with an action of the form

$$S = \int d^3x \left[\frac{k}{4\pi} \epsilon^{\mu\nu\alpha} a_\mu \partial_\nu a_\alpha + a_\mu \left(j^\mu - \frac{e}{2\pi} \epsilon^{\mu\nu\alpha} \partial_\nu A_\alpha \right) \right] \quad (9.23)$$

where a_μ ($\mu = 0, 1, 2$) is a new auxiliary field and j^μ denotes the hole current. The value of the constant k will be specified shortly. Also A_μ is the electromagnetic vector potential. (We are using a three-dimensional covariant notation now. $B^0 = \epsilon^{0ij} \partial_i A_j$ is the magnetic field along the x^3-axis.) The variation of the action with respect to A_α identifies the electromagnetic current as

$$J^\alpha = -\frac{e}{2\pi} \epsilon^{\alpha\mu\nu} \partial_\mu a_\nu \quad (9.24)$$

The equation of motion for the auxiliary field a_μ is

$$\frac{k}{2\pi} \epsilon^{\mu\nu\alpha} \partial_\nu a_\alpha + j^\mu - \frac{e}{2\pi} \epsilon^{\mu\nu\alpha} \partial_\nu A_\alpha = 0 \quad (9.25)$$

From (9.24) and this equation, we see that

$$J^\mu = \frac{e}{k} j^\mu - \frac{e^2}{2\pi k} \epsilon^{\mu\nu\alpha} \partial_\nu A_\alpha. \quad (9.26)$$

Choosing $k = 2p + 1$ we see that we can reproduce the Hall conductivity in (9.21) correctly in the absence of holes. The first term then shows that the charge per hole is e/k.

For a pair of well-separated holes we can take

$$j^\mu = \dot{w}_1^\mu \, \delta^{(2)}(x - w_1) + \dot{w}_2^\mu \, \delta^{(2)}(x - w_2) \quad (9.27)$$

Leaving the electromagnetic field aside for the moment and focusing on the holes, the action becomes

$$S_{\text{hole}} = \frac{k}{4\pi} \int d^3x \; \epsilon^{\mu\nu\alpha} a_\mu \partial_\nu a_\alpha$$
$$+ \int dt \left(a_\mu(w_1) \dot{w}_1^\mu + a_\mu(w_2) \dot{w}_2^\mu + \frac{m \dot{w}_1^2}{2} + \frac{m \dot{w}_2^2}{2} \right) \quad (9.28)$$

where we have also added a regular kinetic energy term for the holes. (The specific form of this will not be important for our purpose.) The time-component of the equation of motion for a_μ, namely (9.25), can be simplified as

$$\partial_z a_{\bar{z}} - \partial_{\bar{z}} a_z = -i \frac{\pi}{k} \left(\delta^{(2)}(x - w_1) + \delta^{(2)}(x - w_2) \right) \quad (9.29)$$

where $z = x^1 + ix^2$, etc. Using the identity

$$\partial_z \frac{1}{\bar{z} - \bar{w}} = \partial_{\bar{z}} \frac{1}{z - w} = \pi \, \delta^{(2)}(x - w) \tag{9.30}$$

the solution to (9.29) can be worked out as

$$a_{\bar{z}} = 0, \qquad a_z = \frac{i}{k} \left(\frac{1}{z - w_1} + \frac{1}{z - w_2} \right) \tag{9.31}$$

The coincident point $w_1 = w_2$ has to be excluded for consistency. We will also use the $a_0 = 0$ gauge; the action (9.28) for the dynamics of the holes can be taken as

$$S = \int dt \left[\frac{m}{2} (\dot{\bar{w}}_1 \dot{w}_1 + \dot{\bar{w}}_2 \dot{w}_2) + a_{w_1} \dot{w}_1 + a_{\bar{w}_1} \dot{\bar{w}}_1 + a_{w_2} \dot{w}_2 + a_{\bar{w}_2} \dot{\bar{w}}_2 \right] \tag{9.32}$$

where we have removed the singularities at the poles; thus in (9.32),

$$a_{w_1} = \frac{i}{k} \frac{1}{w_1 - w_2} + b_{w_1}, \qquad a_{\bar{w}_1} = b_{\bar{w}_1}$$

$$a_{w_2} = \frac{i}{k} \frac{1}{w_2 - w_1} + b_{w_2}, \qquad a_{\bar{w}_2} = b_{\bar{w}_2} \tag{9.33}$$

where b_i are the quantum operators for the gauge fields. The two Eqs. (9.32) and (9.33) suffice for our semiclassical consideration of the statistics of holes (for which b_i can be neglected).[2]

The configuration space for the dynamics of the two holes is $\mathbb{R}^2 \times \mathbb{R}^2 -$ {coincident points}. Because the coincident points $w_1 = w_2$ have been excluded, the closed path of one hole going around the other is not smoothly deformable to zero. In other words, Π_1 of the configuration space is nonzero, equal to $\mathbb{Z}$. In fact, with w_2 fixed,

$$a_{w_1} dw_1 + a_{\bar{w}_1} d\bar{w}_1 = d \left[\frac{i}{k} \log (w_1 - w_2) \right] \tag{9.34}$$

This is evidently closed, a is thus a flat connection on the configuration space. But a cannot be considered exact since

$$\oint_C a = -\frac{2\pi}{k} \neq 0 \tag{9.35}$$

where C is a contour enclosing w_2.

The Hamiltonian corresponding to the action (9.32) is

---

[2] If we quantize the gauge field, (9.29) is the Gauss law. Solving for wave functions (in the $a_{\bar{z}}$-diagonal polarization as in Chap. 7) leads to the same result for $a_w, a_{\bar{w}}$ in the operators in (9.38).

$$H = \frac{1}{2}m\left(\dot{\bar{w}}_1\dot{w}_1 + \dot{\bar{w}}_2\dot{w}_2\right) \tag{9.36}$$

The canonical one-form corresponding to the action (9.32) is

$$\mathcal{A} = \frac{m}{2}\left(\dot{\bar{w}}_1 dw_1 + \dot{w}_1 d\bar{w}_1 + \dot{\bar{w}}_2 dw_2 + \dot{w}_2 d\bar{w}_2\right) + a(1) + a(2)$$
$$a(1) = a_{w_1}dw_1 + a_{\bar{w}_1}d\bar{w}_1$$
$$a(2) = a_{w_2}dw_2 + a_{\bar{w}_2}d\bar{w}_2 \tag{9.37}$$

This has the structure of the usual symplectic potential for particles plus a flat connection, in accordance with the general considerations in Chap. 5. From $\mathcal{A}$, or directly from the action, we can also identify the operators

$$\tfrac{1}{2}m\dot{w}_1 = -i\frac{\partial}{\partial\bar{w}_1} - a_{\bar{w}_1}, \qquad \tfrac{1}{2}m\dot{\bar{w}}_1 = -i\frac{\partial}{\partial w_1} - a_{w_1}$$
$$\tfrac{1}{2}m\dot{w}_2 = -i\frac{\partial}{\partial\bar{w}_2} - a_{\bar{w}_2}, \qquad \tfrac{1}{2}m\dot{\bar{w}}_2 = -i\frac{\partial}{\partial w_2} - a_{w_2} \tag{9.38}$$

This shows that, written as a differential operator, the Hamiltonian will involve the a's. Because of this, it is convenient to write the wave function as

$$\Psi(x_1, x_2) = \exp\left[-\frac{1}{k}\log(w_1 - w_2)\right]\Phi(x_1, x_2)$$
$$= e^{i\Lambda(x_1, x_2)}\Phi(x_1, x_2) \tag{9.39}$$

The action of H on Φ is then the usual one,

$$H\Phi = -\frac{2}{m}\left(\frac{\partial}{\partial w_1}\frac{\partial}{\partial\bar{w}_1} + \frac{\partial}{\partial w_2}\frac{\partial}{\partial\bar{w}_2}\right)\Phi \tag{9.40}$$

The use of $\Phi(x_1, x_2)$ as in (9.39) maps the problem to one where the connections a, $\bar{a}$ do not appear in the Schrödinger equation.

We can now consider the exchange of the two holes as due to a rotation of the two points by π followed by a translation to bring them back to the same points. Since Φ is the wave function on which H acts without any extra flat connection a, we can take Φ to be symmetric under exchange. As for the prefactor $e^{i\Lambda}$ in (9.39), the translation does not change it, but the π-rotation leads to

$$e^{i\Lambda(x_1, x_2)} = e^{-i\pi/k}e^{i\Lambda(x_1, x_2)} \tag{9.41}$$

With $k = 2p + 1$, we see that the two holes do display fractional statistics. The origin of this can be traced to the closed but not exact one-form (9.34), which is itself related to the nontrivial connectivity of the configuration space.

In two spatial dimensions, it is also possible to have fractional values for the spin for a particle [31]. The usual argument for the quantization for spin in three spatial dimensions relies on two ingredients: the fact that the components of the angular momentum operators do not commute among themselves and the requirement of unitarity for the representation. In two spatial dimensions, where there is only one rotation, fractional values for spin are possible. This is true even in a Lorentz-invariant theory, because of the noncompact nature of the Lorentz group [32]. There is a spin-statistics theorem in two spatial dimensions as well. In accordance with this, the result we have shown implies that the holes must also have fractional spin or that they are "anyons" [31].

Another comment which may be of interest is the following. Consider the operators

$$p_k = -i \frac{\partial}{\partial \bar{w}_k} - a_{\bar{w}_k}, \qquad p'_k = -i \frac{\partial}{\partial \bar{w}_k}, \qquad k = 1, 2, \tag{9.42}$$

with similar operators for the derivatives with respect to w_k. Notice that if we exclude the coincident point $w_1 = w_2$, then $a(1)$ and $a(2)$ are flat connections by virtue of (9.29). As a result, both sets of operators $\{p_k, \bar{w}_k, \bar{p}_k, w_k\}$ and $\{p'_k, \bar{w}_k, \bar{p}'_k, w_k\}$ obey the Heisenberg algebra. Formally, they are related by the transformation,

$$e^{-i\Lambda} p_k e^{i\Lambda} = p'_k \tag{9.43}$$

However, even the pure phase part of this transformation (i.e., $e^{i(\Lambda + \bar{\Lambda})/2}$) is not a unitary transformation since it is not single-valued on the space. Therefore, the two sets $\{p_k, \bar{w}_k, \bar{p}_k, w_k\}$ and $\{p'_k, \bar{w}_k, \bar{p}'_k, w_k\}$ constitute inequivalent representations of the Heisenberg algebra. The Stone-von Neumann theorem tells us that a finite number of Heisenberg algebras have unique representations (up to unitary equivalence) if the underlying space is simply connected, but that they can have inequivalent representations if the space is not simply connected. Since the coincident point $w_1 = w_2$ has been excluded from consideration, the position space for the quasiparticles is not simply connected. So the result we find can be viewed as exemplifying this particular feature of the Stone-von Neumann theorem.

Problem

9.1 Obtain the Landau levels for electrons in a uniform Abelian magnetic field on $\mathbb{CP}^2$ and show that the lowest level wave functions agree with (6.70).

Open Access This chapter is licensed under the terms of the Creative Commons Attribution 4.0 International License (http://creativecommons.org/licenses/by/4.0/), which permits use, sharing, adaptation, distribution and reproduction in any medium or format, as long as you give appropriate credit to the original author(s) and the source, provide a link to the Creative Commons license and indicate if changes were made.

The images or other third party material in this chapter are included in the chapter's Creative Commons license, unless indicated otherwise in a credit line to the material. If material is not included in the chapter's Creative Commons license and your intended use is not permitted by statutory regulation or exceeds the permitted use, you will need to obtain permission directly from the copyright holder.

Chapter 10
Fluid Dynamics

We now turn to considerations about how an action for fluid dynamics can be constructed using the results o the geometric quantization of G/H spaces and how anomalies can be incorporated into fluid dynamics [33]. Co-adjoint orbit actions, introduced in Chap. 6, will be used to set up a group-theoretic formulation of fluid dynamics. We will start with the well known formulation of classical nonrelativistic fluid dynamics.

10.1 The Lagrange Formulation

The so-called Lagrange formulation of fluid dynamics, developed more than two centuries ago by Euler and Lagrange, is an elegant method of obtaining the equations of fluid dynamics starting from Newton's equations for point-particles. Here one considers a collection of, say, N particles obeying the equations of motion

$$\frac{\mathrm{d}}{\mathrm{d}t}\dot{X}^i_\lambda = -\frac{\partial V}{\partial X_{i\lambda}} \tag{10.1}$$

where X^i_λ denote the position of the λ-th particle, $\lambda = 1, 2, \ldots, N$. For simplicity, we have taken all particles to have the same mass m, with units adjusted so that $m = 1$. We can label the particles by their positions at time $t = 0$, assuming that there is no overlap of particles. In this way of labeling particles, λ is a three-vector corresponding to the initial position vector. In the limit of a large number of particles, we may take λ to be continuous. Let $\rho_0(\lambda)$ be the number density of particles. Then we sum Eq. (10.1) over a small range of λ and go to the continuous λ-limit to obtain

$$\rho_0(\lambda)\,\mathrm{d}^3\lambda\,\frac{\mathrm{d}}{\mathrm{d}t}\dot{X}^i(t, \lambda) = -\rho_0(\lambda)\,\mathrm{d}^3\lambda\,\frac{\partial V}{\partial X_i(t, \lambda)} \tag{10.2}$$

© The Author(s) 2024
V. P. Nair, *Geometric Quantization and Applications to Fields and Fluids*,
SpringerBriefs in Physics, https://doi.org/10.1007/978-3-031-65801-3_10

Now, the particle at position $\boldsymbol{\lambda}$ at $t = 0$ moves to $X^i(t, \lambda)$ at time t. This is a continuous transformation of the λ^i into X^i, which is invertible at least for small t. We can therefore solve for λ^i as a function of X^i and t and write various quantities as functions of t, X^i. Since the number of particles is conserved, we should have $\rho_0(\lambda)\, d^3\lambda = \rho(t, X)\, d^3X$. This shows that we can define the density of particles in terms of X as

$$\rho(t, X) = \frac{\rho_0(\lambda)}{\det(\partial X/\partial \lambda)} \tag{10.3}$$

The density ρ so defined obeys an equation of continuity. By direct differentiation with respect to time, we find

$$\frac{\partial \rho}{\partial t} + \frac{\partial \rho}{\partial X^i}\dot{X}^i = \rho_0(\lambda)\frac{d}{dt}\frac{1}{\det(\partial X/\partial \lambda)} = -\frac{\rho_0(\lambda)}{\det(\partial X/\partial \lambda)}\frac{d}{dt}(\log\det(\partial X/\partial \lambda))$$

$$= -\rho\,\frac{\partial \lambda^i}{\partial X^k}\frac{\partial \dot{X}^k}{\partial \lambda^i} = -\rho\,\nabla_k \dot{X}^k \tag{10.4}$$

We now define the velocity at a point X as

$$v^i(t, X) = \dot{X}^i(t, \lambda)\Big]_{\lambda=\lambda(t,X)} \tag{10.5}$$

Equation (10.4) then reduces to the continuity equation

$$\frac{\partial \rho}{\partial t} + \nabla_k(\rho v^k) = 0 \tag{10.6}$$

The equation of motion (10.2) involves the time-derivative of the velocity $\dot{X}^i(t, \lambda)$ at fixed λ. If we substitute for λ in terms of X in $\dot{X}^i$, i.e., use $\dot{X}^i(t, \lambda(t, X))$, we have

$$\frac{d\dot{X}^i}{dt} = \frac{d\dot{X}^i}{dt}\Big]_{\lambda \text{ fixed}} + \frac{\partial \dot{X}^i}{\partial \lambda^k}\dot{\lambda}^k \tag{10.7}$$

Since λ^i (being the initial positions) do not depend on time, being initial data, we also have the identity

$$0 = \frac{\partial \lambda^i(t, X)}{\partial t} + \frac{\partial \lambda^i(t, X)}{\partial X^k}\dot{X}^k \tag{10.8}$$

Using this in (10.7) we get

$$\frac{d\dot{X}^i}{dt}\Big]_{\lambda \text{ fixed}} = \frac{d\dot{X}^i}{dt} + \frac{\partial \dot{X}^i}{\partial \lambda^k}\frac{\partial \lambda^k}{\partial X^l}\dot{X}^l$$

$$= \frac{dv^i}{dt} + v^k\nabla_k v^i \tag{10.9}$$

Thus the equation of motion (10.2) becomes

$$\rho\left[\frac{\partial v^i}{\partial t} + v^k \nabla_k v^i\right] = -\rho \frac{\partial V}{\partial X_i} \tag{10.10}$$

This equation, which is known as the Euler equation, along with the continuity Eq. (10.6), defines perfect fluid dynamics. The right hand side of (10.10) can be expressed in terms of the gradient of the pressure, but we will not need that for now.

In modern physics, a point-particle is defined as a unitary irreducible representation (UIR) of the Poincaré group which is the group of spacetime translations and Lorentz transformations. In addition, we may want to consider particles with internal symmetries such as nonabelian color charges, the latter being also described by an appropriate representation of the symmetry group. So we may ask:

> Can we do a Lagrange trick and describe fluid dynamics in terms of group theory, with each particle corresponding to a unitary irreducible representation of the symmetry group (Poincaré $\otimes$ internal symmetry group)?

Beyond the formalistic value, there are some good reasons why this will be of interest. In such a formalism, symmetry would be really foundational and this would facilitate the inclusion of nonabelian internal symmetries and spin in magnetohydrodynamics and also incorporate anomalous symmetries as well. These have all become issues of interest in recent research partly because of the deconfined fluid phase of quarks and gluons.

We start with a simple case of a nonrelativistic particle which carries an internal symmetry, say, $SU(2)$ to see how this can all work out. (This internal symmetry could be "color" or spin or something else depending on the physical context.) The action for such a particle coupled to an $SU(2)$ gauge field is given by

$$\begin{aligned}
S &= \int dt \left[\frac{1}{2} m \dot{x}^2 - A_i^a Q^a \dot{x}_i - i \frac{n}{2} \operatorname{Tr}(\sigma_3 g^{-1} \dot{g})\right] \\
&= \int dt \left[\frac{1}{2} m \dot{x}^2 - i \frac{n}{2} \operatorname{Tr}(\sigma_3 g^{-1} D_0 g)\right]
\end{aligned} \tag{10.11}$$

where $Q^a = \frac{n}{4} \operatorname{Tr}(\sigma_3 g^{-1} \sigma^a g)$ and $D_0 = \partial_0 + A_i^a \dot{x}_i (-i\sigma^a/2)$. D_0 is the covariant derivative of g with respect to the $SU(2)$ gauge field evaluated on the trajectory of the particle. This action was proposed in the 1970s by Balachandran and collaborators [11]; the equations of motion corresponding to this action were written down earlier, in 1971, by Wong [10]. The last term in (10.11), apart from the gauge field term, is familiar to us as the action (6.42) for $G/H = SU(2)/U(1)$. The quantization of the action is also familiar. The usual kinetic term $\frac{1}{2} m \dot{x}^2$ will lead to the usual point particle dynamics, with a minimal coupling to the gauge field via the charge operator Q^a. The degrees of freedom represented by g will lead to a unitary representation of $SU(2)$, with $j = \frac{n}{2}$. This part will describe the dynamics of the internal symmetry and how it influences and is influenced by the kinetic motion of the particle and by the external field.

We can now see how to generalize to fluids. We will focus on the last term in (10.11) as it is the key term for obtaining UIRs of the group after quantization. We consider a large number of particles, using a variable λ to label them. As with the Lagrangian approach to fluids, we will eventually take λ to be continuous and to correspond to a three-volume. For the last term in (10.11) we get

$$
S = -i\frac{n}{2} \int dt\, \mathrm{Tr}(\sigma_3\, g^{-1}\dot{g}) \longrightarrow S = -\frac{i}{2} \int dt \sum_\lambda n_\lambda \mathrm{Tr}(\sigma_3\, g_\lambda^{-1}\dot{g}_\lambda) \qquad (10.12)
$$

We can take the continuum limit by $\sum_\lambda \to \int J\, d^3x/v$, where J is the Jacobian of the transformation $\boldsymbol{\lambda} \to \boldsymbol{x}$, $J = |\partial\lambda/\partial x|$ and v indicates a small volume over which the dynamics is coarse-grained. Defining a density by $j^0 = n\, J/v$, we get [34]

$$
S = -i \int d^4x\, j^0\, \mathrm{Tr}(t_3 g^{-1}\partial_0 g), \qquad t_3 = \frac{\sigma_3}{2} \qquad (10.13)
$$

where $g(t, \lambda) = g(t, \boldsymbol{x})$ is to be considered as a spacetime-dependent group element. The form of the action (10.13) also suggests a natural relativistic generalization

$$
S = -i \int j^\mu\, \mathrm{Tr}(t_3\, g^{-1}\partial_\mu g) \qquad (10.14)
$$

The remaining terms in the action can be added on at this stage, but before doing that, we pause to consider what happens with the Poincaré group. If we follow the same strategy we should consider the analog of the term $\mathrm{Tr}(t_3\, g^{-1}\dot{g})$ for the Poincaré group, which has the translational parameters x^μ and the rotational and Lorentz boost parameters; the latter set of parameters may be gathered into a Lorentz group element Λ. The action is then given by

$$
S = -\int d\tau\, p_\mu \dot{x}^\mu + i\frac{n}{4} \int d\tau\, \mathrm{Tr}(\Sigma_3\, \Lambda^{-1}\dot{\Lambda}) \qquad \Sigma_3 = \begin{bmatrix} \sigma_3 & 0 \\ 0 & \sigma_3 \end{bmatrix} \qquad (10.15)
$$

where we have chosen to display the term involving the Lorentz group element Λ in terms of the usual spinor representation.[1] This is almost what we want, but the first term in the action (10.15) needs some rewriting. This is because, in going over to a fluid description, the position variables x^μ are a bit awkward. First of all, there should only be three independent x's or corresponding velocities. For the point-particle, this is naturally implemented by a mass-shell type constraint. It is not clear how to do this for fluids. Secondly, the role of diffeomorphisms versus translations is not clear in this language. So we will first deal with this problem before returning to the main line of development.

---

[1] This is like using the 2×2-matrix version of g to display the action (10.11). It does not imply that there is anything special about this representation.

10.2 Clebsch Variables and the General Form of Action

We return to the usual approach to fluids briefly. It has been known for a long time that fluid dynamics can be described as a Poisson bracket system. This means that the equations of motion are derived from a Hamiltonian

$$H = \int d^3x \left[\frac{1}{2} \rho\, v^2 + V(\rho) \right] \tag{10.16}$$

by using the Poisson bracket

$$\{F, G\} = \int \left[-\frac{\delta F}{\delta \rho} \partial_i \left(\frac{\delta G}{\delta v_i} \right) + \frac{\delta G}{\delta \rho} \partial_i \left(\frac{\delta F}{\delta v_i} \right) + \frac{\omega_{ij}}{\rho} \frac{\delta F}{\delta v_i} \frac{\delta G}{\delta v_j} \right] \tag{10.17}$$

We have written the form of the bracket for arbitrary functions F, G of the fluid variables and $\omega_{ij} = \partial_i v_j - \partial_j v_i$ is the vorticity. The potential energy term in H, namely $V(\rho)$, is related to the pressure of the fluid as $P = \rho \frac{\partial V}{\partial \rho} - V$.

It is then easy to check that any local observable F will Poisson commute with the helicity which is defined as

$$C = \frac{1}{12\pi^2} \int \epsilon^{ijk}\, v_i\, \partial_j v_k \tag{10.18}$$

where we take the velocity to vanish at the boundary of the spatial region of integration. Denoting the variables ρ, v_i collectively as q^μ, and writing the Poisson brackets as $\{q^\mu, q^\nu\} = K^{\mu\nu}$, we can check from (10.17) that $\delta C/\delta v_i$ is a zero mode for $K^{\mu\nu}$. This means that $K^{\mu\nu}$ is not invertible. Comparing with (2.12) we see that we have a problem. If $K^{\mu\nu}$ has an inverse, that would be the symplectic structure $\Omega_{\mu\nu}$ and we can construct an action. But that is not possible because $K^{\mu\nu}$ has a zero mode. This is a problem, but the way to a solution is also clear. Since C Poisson commutes with any local observable, it must be superselected. We must fix its value and then consider only those velocities which keep the value unchanged. Such a parametrization is given by the Clebsch variables which expresses the velocity as

$$v_i = \partial_i \theta + \alpha\, \partial_i \beta \tag{10.19}$$

where θ, α and β are 3 independent fields. One can easily check that the integrand of C is a total derivative with this parametrization and gives zero upon integration. (We can also accommodate other values of C, see below.) A suitable action which gives the fluid equations is then

$$S = \int d^4x \left[\rho\, \dot\theta + \rho \alpha\, \dot\beta \right] - \int d^4x \left[\frac{1}{2} \rho v^2 + V \right] \tag{10.20}$$

We can also write this as

$$S = \int d^4x \left[j^\mu \left(\partial_\mu \theta + \alpha \, \partial_\mu \beta \right) \right] - \int d^4x \left[j^0 - \frac{j^i j^i}{2\rho} + V \right] \qquad (10.21)$$

where $j^0 = \rho$ and we introduce an auxiliary field $\mathbf{j}$. Elimination of $\mathbf{j}$ takes us back to (10.20). This is easily generalized to the relativistic case as

$$S = \int d^4x \left[j^\mu \left(\partial_\mu \theta + \alpha \, \partial_\mu \beta \right) - F(n) \right] \qquad (10.22)$$

where $F(n) = n + V(n)$ and $n^2 = j^2 = (j^0)^2 - j^i j^i$. Notice that $n = \sqrt{(j^0)^2 - \mathbf{j} \cdot \mathbf{j}} \approx j^0 - (j^i j^i / 2 j^0) + \cdots$, so that (10.21) is recovered in the nonrelativistic case.

The Clebsch parametrization can also be written in a group-theoretic form [35, 36]. For this purpose, we can use either $SU(1, 1)$ or $SU(2)$. We parametrize an element of the group as[2]

$$g = \frac{1}{\sqrt{1 \mp \bar{u}u}} \begin{pmatrix} 1 & u \\ \pm \bar{u} & 1 \end{pmatrix} \begin{pmatrix} e^{i\frac{\theta}{2}} & 0 \\ 0 & e^{-i\frac{\theta}{2}} \end{pmatrix}, \qquad (10.23)$$

We can easily check that

$$- i \, \mathrm{Tr} \left(\sigma_3 \, g^{-1} dg \right) = d\theta + \alpha \, d\beta, \qquad \alpha = \frac{\bar{u}u}{(1 \mp \bar{u}u)}, \qquad \beta = \mp i \log(u/\bar{u}) \quad (10.24)$$

where the upper sign applies to $SU(1, 1)$ and the lower to $SU(2)$.[3] We can now write the usual ordinary fluid dynamics action as

$$S = \int d^4x \left[-i j^\mu \, \mathrm{Tr}(\sigma_3 \, g^{-1} \partial_\mu g) - F(n) \right] \qquad (10.25)$$

We have thus brought the action, even for the usual fluid dynamics, to a form consistent with the group-theoretic approach. We can now see how the Poincaré group can be accommodated. For the translational part we use the action in terms of the Clebsch variables. For the rest of it, we can use the usual group-theoretic way which we have already discussed. The action for general fluid dynamics is thus given by [37]

---

[2] Whether we should choose $SU(1, 1)$ or $SU(2)$ depends on the vorticity which is given as $d\alpha \, d\beta$. The group $SU(2)$ would describe situations with quantized vorticity, $SU(1, 1)$ would give no quantization condition on vorticity.

[3] By the way, we are also saying that ordinary fluid dynamics can display an $SU(1, 1)$ or $SU(2)$ symmetry, which is effectively replacing the diffeomorphism symmetry. This is a point worth further exploration.

$$\mathcal{S} = \int \mathrm{d}^4 x \left[j^\mu \left(\partial_\mu \theta + \alpha \, \partial_\mu \beta \right) - \frac{i}{2} \, j^\mu_{(s)} \, \mathrm{Tr} (\Sigma_3 \, \Lambda^{-1} \partial_\mu \Lambda) \right.$$
$$\left. - i \sum_a j^\mu_a \mathrm{Tr} (q_a \, g^{-1} D_\mu \, g) - F(\{n\})) \right] + S(A) \qquad (10.26)$$

We use q_a to denote the diagonal generators of the internal symmetry group G with $g \in G$. The currents j^μ, $j^\mu_{(s)}$, j^μ_a correspond to the transport of mass, spin and internal quantum numbers, respectively. Generally, we must have different currents j^μ, $j^\mu_{(s)}$, j^μ_a for mass flow, spin flow and the transport of other quantum numbers, since they are independent. For example, we may have a cluster of particles of zero total spin moving off in some direction, giving mass transport but no spin transport; we can have a similar situation with internal symmetry groups as well. Generally these currents are independent; any relations among them must be viewed as "constitutive relations" characteristic of the physical system. The coupling of the system to gauge fields follows from covariant derivatives on the group elements. The function $F(\{n\})$ depends on all the invariants such as $n = \sqrt{j^\mu j_\mu}$, $n_a = \sqrt{j^\mu_a j_{\mu a}}$, $n_{ab} = \sqrt{j^\mu_a j_{\mu b}}$ etc., which we can make from the currents and the gauge fields. We have explicitly indicated the action for the gauge fields. The group-valued fields are related to flow velocities and currents and are given by the equations of motion,

$$\frac{1}{n} \frac{\partial F}{\partial n} \, j_\mu = \partial_\mu \theta + \alpha \, \partial_\mu \beta$$
$$\frac{1}{n_a} \frac{\partial F}{\partial n_a} \, j_{\mu a} = -i \, \mathrm{Tr} \, (q_a \, g^{-1} D_\mu \, g), \qquad \text{etc.} \qquad (10.27)$$

10.3 Assorted Comments

Many new concepts (or at least concepts which may not be very familiar) have been introduced, so a few clarifying remarks are in order at this point.

Helicity

In terms of the group-valued variables, the helicity is given by the topological invariant

$$C = \frac{1}{24\pi^2} \int \mathrm{Tr} (g^{-1} \mathrm{d}g)^3 \qquad (10.28)$$

This shows how we may generalize the Clebsch parametrization to situations with nonzero value of C. We choose a particular g, say g_1 which gives the desired value C. Then we use $g_1 \, g$ in place of g in (10.24) to get the parametrization for velocities. g is taken to have zero C. It is easy to check that $C[g_1 \, g] = C[g_1] + C[g] = C[g_1]$. Notice also that $C[g]$ is basically $-Q[g]$, the winding number in (7.52), for the case of $g \in SU(2)$. For $g \in SL(2, \mathbb{R})$, we still have the expression (10.28) for C, but the result does not yield a quantized number since $\Pi_3(SL(2, \mathbb{R})) = 0$.

The action and the density matrix

The idea of using an action of the form $\int j^0 \operatorname{Tr}(\sigma_3 g^{-1} \dot{g})$ may be seen from another more general point of view as well. The full quantum dynamics for a state with density matrix ρ is given by the action

$$S = \int dt \operatorname{Tr} \left[\rho_0 \left(U^\dagger i \frac{\partial U}{\partial t} - U^\dagger H U \right) \right] \tag{10.29}$$

where U is a general unitary transformation. The variational equation for this is

$$i \frac{\partial \rho}{\partial t} = H \rho - \rho H, \qquad \rho = U \rho_0 U^\dagger \tag{10.30}$$

which is the expected equation for the time-evolution of the density matrix. The canonical one-form corresponding to this action is

$$\mathcal{A} = i \operatorname{Tr}(\rho_0 U^\dagger \delta U) \tag{10.31}$$

where δU includes all possible observables. Suppose we now restrict ourselves to the dynamics of a smaller number of observables, say those corresponding to symmetry transformations which can survive into the hydrodynamic regime. Let these symmetry transformations form a Lie group G. Then for $g \in G$, parametrized in terms of θ^A, we can write

$$g^{-1} dg = -i t_a \mathcal{E}_A^a d\theta^A \tag{10.32}$$

where t_a are the generators of the group and this equation defines the one-forms $\mathcal{E}_A^a d\theta^A$. For the variation of unitary transformations corresponding to this subset in the quantum theory, we can then write

$$U^\dagger \delta U = -i \hat{q}_a \mathcal{E}_A^a \delta\theta^A \tag{10.33}$$

where $\hat{q}_a$ are the quantum operators corresponding to the generators of the group. This shows that $\mathcal{A}$ restricted to the variables of interest is

$$\mathcal{A} = 2i \rho_a \operatorname{Tr}(t_a g^{-1} dg), \qquad \rho_a = \operatorname{Tr}(\rho_0 \hat{q}_a) \tag{10.34}$$

We can now ask the question: What is the action (at the level of the reduced set of observables θ's) which gives this $\mathcal{A}$? This is evidently the co-adjoint orbit action of the form we have been using. The variables θ's are essentially the relevant collective variables of the theory in the regime of interest.

Diffeomorphisms and Clebsch variables

Finally, we can think of the Clebsch variables in another way as well. We start by looking at the diffeomorphism algebra,

$$\{M(\xi), M(\xi')\} = M(\xi \times \xi'), \qquad (\xi \times \xi')^i = \xi^k \partial_k \xi'^i - \xi'^k \partial_k \xi^i \qquad (10.35)$$

where M is the generator of spatial diffeomorphisms, given by T_{0i} where $T_{\mu\nu}$ is the energy-momentum tensor. The algebra (10.35) can be realized by

$$J_i = \pi_1 \, \partial_i \varphi_1 + \pi_2 \, \partial_i \varphi_2 + \cdots \qquad (10.36)$$

for any number of canonical pairs of variables (π_i, φ_i). We need two such pairs for a complete characterization in 3 spatial dimensions. Hence, we can see that diffeomorphism symmetry can be traded for an $SU(1, 1)$ or $SU(2)$ symmetry for the pairs π_i, φ_i. The redesignation of variables as $\pi_1 = \rho, \pi_2 = \rho \alpha, \varphi_1 = \theta, \varphi_2 = \beta$ takes us back to the usual Clebsch form.[4]

We can also view π_1, φ_1 as the modulus and phase of a complex field ψ, ψ^*. The interpretation of α, β, which will need another complex field, is a little more subtle. Recall that α, β are the fields required to get nonzero vorticity. We may observe that for vorticity, we need to compare the velocities of nearby particles. Thus in attributing some nonzero vorticity to each local coarse-graining unit, we see that inside each such unit (around, say, x), we must have distinct fields representing these particles whose velocities are to be compared. This means that $\psi(x)$ and $\psi(x + \epsilon)$ must be counted as independent fields since we want to replace them by fields at a single point x upon coarse-graining. This gives some understanding of how the $SU(1, 1)$ or $SU(2)$ group emerges.

10.4 Examples

10.4.1 Nonabelian Magnetohydrodynamics

We will briefly mention a few examples before going on to the question of anomalies. Our first example is about nonabelian magnetohydrodynamics, say with $SU(2)$ as the internal symmetry [34]. Picking out the relevant terms in the general action (10.26), we see that we can take the action for this case as

$$S = \int j^\mu_{(m)} \left(\partial_\mu \theta + \alpha \, \partial_\mu \beta\right) - i \int j^\mu \, \mathrm{Tr}(\sigma_3 \, g^{-1} D_\mu g) - \int F(n)$$
$$+ S_{\mathrm{YM}} \qquad (10.37)$$

---

[4] In principle, we can use the action (10.15) with the translational degrees of freedom x^μ even in the fluid case, instead of the Clebsch variables. If we keep $\dot{x}^\mu$ as fluid velocity, then we do get the correct fluid equations, but with no pressure.

As a reminder, our conventions in this expression are as follows.

$$D_\mu g = \partial_\mu g + A_\mu\, g, \qquad j^\mu_{(m)} = n_m\, U^\mu, \qquad j^\mu = n\, u^\mu \tag{10.38}$$

Also, $A_\mu = -i\, t^a\, A^a_\mu$, $t^a = \frac{1}{2}\sigma^a$. $j^\mu_{(m)}$ denotes the mass current, while j^μ corresponds to the current for the diagonal generator of the internal symmetry. We have also defined the flow velocities U^μ and u^μ in terms of the currents; they are 4-vectors normalized to unity, $U^2 = 1$, $u^2 = 1$.

The current which couples to the gauge field may be obtained as

$$J^\mu_a = -\frac{\delta S}{\delta A^a_\mu} = \mathrm{Tr}(\sigma_3\, g^{-1} t_a g)\, j^\mu = Q_a\, u^\mu$$

$$Q_a = n\, \mathrm{Tr}(\sigma_3\, g^{-1} t_a g) \tag{10.39}$$

Notice that the current factorizes into a charge density Q_a and a flow velocity u^μ. This is known as the Eckart factorization. The equations of motion may be derived from the action (10.37) by varying all the fields. We show some of the equations here:

$$\partial_\mu j^\mu = 0$$
$$(D_\mu J^\mu)_a = 0$$
$$[j^\mu g^{-1} D_\mu g, \sigma_3] = 0$$
$$n\, u^\mu \partial_\mu(u_\nu F') - n\, \partial_\nu F' = -J^\mu_a F^a_{\mu\nu} \tag{10.40}$$

The first of these equations arises from right transformations of the form $g \to g(1 + \sigma_3\epsilon)$, the second from general left transformations of g. The third equation is from arbitrary right translations. The last equation is obtained by applying $u^\mu\partial_\mu$ to the equation resulting from variations of j^μ. The first two are conservation laws, while the last one is the Euler equation for the (nonabelian) charge transport.[5] The first three equations in (10.40) also give

$$u^\mu(D_\mu Q)_a = (D_0 Q)_a + \boldsymbol{u} \cdot (\boldsymbol{D} Q)_a = 0 \tag{10.41}$$

This may be viewed as the fluid version of the Wong equations for the transport of nonabelian charge by a point-particle. We also have $\partial_\mu T^{\mu\nu} = \mathrm{Tr}\,(J^\mu F_{\mu\nu})$ where the energy-momentum tensor $T_{\mu\nu}$ has the perfect fluid form.

The group element g may be given a nice physical interpretation. The nonabelian charge density $\rho = \rho_a\, t_a$ (which is the time-component of J^μ_a) transforms, under gauge transformations, as

$$\rho \to \rho' = h^{-1}\rho\, h, \qquad h \in SU(2) \tag{10.42}$$

---

[5] There is another equation for mass transport which we are not displaying. Here we are zeroing in on just the "new" equations, namely, those beyond the usual ones from variation of θ, α, β, etc.

Thus we can diagonalize ρ at each point by an (x, t)-dependent transformation g. Then we can write $\rho = g \, \rho_{\text{diag}} \, g^{-1}$, with $\rho_{\text{diag}} = \rho_0 \sigma_3$. In other words,

$$\rho_a = \rho_0 \, \text{Tr}(g \, \sigma_3 \, g^{-1} t_a) = j^0 \, \text{Tr}(g \, \sigma_3 \, g^{-1} t_a) \tag{10.43}$$

The group element g diagonalizes the charge density at each point. The eigenvalues are gauge-invariant and are represented by n. We may thus view g as describing the degrees of freedom corresponding to the orientation of the local charge density in color space. Under a gauge transformation, $g \to h^{-1} g$.

The Poisson brackets involving the charge densities are

$$\{j^0(x), j^0(y)\} = 0$$
$$\{j^0(x), g(y)\} = -i \, g(x) \left(\frac{\sigma_3}{2}\right) \delta(x - y)$$
$$\{J_a^0(x), J_b^0(y)\} = f_{abc} \, J_c^0(x) \, \delta(x - y)$$
$$\{J_a^0(x), g(y)\} = -i \left(\frac{\sigma_a}{2}\right) g(x) \, \delta(x - y) \tag{10.44}$$

Notice that J_a^0 generates left transformations on g, while j^0 generates right transformations along the σ_3-direction.

10.4.2 Spin and Fluids

Another example we will briefly quote is for fluids with spin [37]. Consider a special case where mass transport and charge transport are described by the same flow velocity. In other words, impose a "constitutive relation" $(e/m) j_{(m)}^\mu = j_e^\mu$. Such a relation is reasonable when we have one species of particles with the same charge. Further, for dilute systems, if we neglect the possibility of spin-singlets forming (and moving independently), we can take spin flow velocity $\approx$ charge flow velocity, so that we can further impose $(s/m) j_{(m)}^\mu = j_{(s)}^\mu$.[6] (We will set $m = 1$ from now on, for simplicity, by a proper choice of units.) In this case, the action (10.26) simplifies, with a single current j^μ, as

$$S = S(A) + \int d^4x \left[j^\mu \left(\partial_\mu \theta + \alpha \partial_\mu \beta + e A_\mu\right) - \frac{i}{4} j^\mu \, \text{Tr}(\Sigma_3 \, \Lambda^{-1} \partial_\mu \Lambda) - F(n, \sigma) \right] \tag{10.45}$$

The Lorentz group element Λ may be written as $\Lambda = B \, R$, where B is a specific boost transformation taking us from a rest frame to a moving frame and R is a spatial rotation. Explicitly,

---

[6] Here we are considering a very special case to illustrate certain physical results. The general formalism allows for the discussion of the case with independent transport of mass, charge and spin.

$$B(u) = \frac{1}{\sqrt{2(u^0+1)}} \begin{bmatrix} u^0+1 & \sigma \cdot u \\ \sigma \cdot u & u^0+1 \end{bmatrix} \qquad (10.46)$$

The statement that $j^\mu_{(m)} = j^\mu_{(s)}$ means that B contains the same velocity u^μ as for the mass transport, as in $j^\mu = n\,u^\mu$. The function F in (10.45) depends on n and $\sigma = S^{\mu\nu} F_{\mu\nu}$, where $S^{\mu\nu}$ is the spin density,

$$S^{\mu\nu} = \frac{1}{2}\,\mathrm{Tr}\,(\Sigma_3\,\Lambda^{-1}\,J^{\mu\nu}\,\Lambda), \qquad J^{\mu\nu} = \frac{i}{4}[\gamma^\mu, \gamma^\nu] \qquad (10.47)$$

Here γ^μ are the Dirac γ-matrices, obeying

$$\gamma^\mu\gamma^\nu + \gamma^\nu\gamma^\mu = 2\,g^{\mu\nu} \qquad (10.48)$$

One interesting feature which emerges from this analysis, and the equations of motion for the action (10.45), is that the spin density is subject to precession effects due to pressure gradient terms in addition to the expected precession due to the magnetic field. This is seen explicitly from the equations of motion

$$u^\alpha \partial_\alpha(F'\,u_\nu) - \partial_\nu F' = e\,u^\lambda\,F_{\lambda\nu} + \cdots$$
$$u^\alpha \partial_\alpha S_{\mu\nu} = \frac{e}{F'}\left[S_\mu^{\ \lambda}F_{\lambda\nu} - S_\nu^{\ \lambda}F_{\lambda\mu}\right] + \left[S_\mu^{\ \lambda} f_{\lambda\nu} - S_\nu^{\ \lambda} f_{\lambda\mu}\right] + \cdots$$
$$(10.49)$$

where $F' = (\partial F/\partial n)$ and

$$f_{\lambda\nu} = \frac{1}{F'}\left[u_\lambda\,\partial_\nu F' - u_\nu\,\partial_\lambda F'\right] \qquad (10.50)$$

The first equation in (10.49), which is the Euler equation, shows the expected Lorentz force formula for charged fluids. The first term on the right hand side of the second equation describes the precession of the spin density in the electromagnetic field. The second term, $S_\mu^{\ \lambda} f_{\lambda\nu} - S_\nu^{\ \lambda} f_{\lambda\mu}$, describes a spin precession effect due to pressure gradient terms which can exist even in the absence of external fields. This is a bit unusual and is a novel effect in the context of fluid dynamics, although its origin can be traced to the spin-orbit couplings in the relativistic theory. There are corrections to both these equations depending on the gradient of the spin density, as indicated by the ellipsis.

10.5   Anomalies in Fluid Dynamics

We will now consider how anomalies can affect fluid dynamics. Anomalies arise in the quantum theory because of the need to regularize the theory. This involves an upper (ultraviolet) cut-off on the integrations over loop momenta in various Feynman diagrams. If a situation arises that one cannot find a regulator which preserves all the classical symmetries, then we have to ensure that the regulator we choose preserves gauge symmetries (for consistency reasons). This may mean that we have to give up some of the other non-gauge symmetries, with the corresponding currents not being conserved. We say that those symmetries are anomalous.

Even though anomalies arise out of ultraviolet regulators, they have a deeper topological origin and one consequence of this aspect of the anomalies is that they are not renormalized. Further, they can also be reproduced from infrared physics. Because of this property, we should expect the anomalies to be present in all phases of the theory. In particular, we can expect them to be relevant in the hydrodynamical regime as well.

10.5.1   Anomalous Electrodynamics

First of all we will consider a very simple case, that of an Abelian $U(1)$ theory which has anomalies. We may think of this as electromagnetism. The basic equations we need are the conservation laws,

$$\partial_\mu T^\mu_{\ \nu} = F_{\nu\mu} J^\mu$$
$$\partial_\mu J^\mu = -\frac{c}{8} \epsilon^{\mu\nu\alpha\beta} F_{\mu\nu} F_{\alpha\beta} \tag{10.51}$$

The first equation is the expected relation for the divergence of the energy-momentum tensor. The second one is the conservation law for charge which is anomalous, with the anomaly as given on the right hand side. Here c is a constant, the anomaly coefficient, which can be calculated from the underlying quantum physics. The lack of conservation for the electric current will, of course, lead to inconsistencies, so we must really regard this system as describing a subsystem which is anomalous, with another subsystem which will cancel this anomaly for the full system, thus avoiding any inconsistencies. These two Eq. (10.51) are to be supplemented by the form of T^μ_ν and J^μ, given by

$$T^\mu_{\ \nu} = \mu\, n\, U^\mu U_\nu + \delta^\mu_{\ \nu}\, P$$
$$J^\mu = n\, U^\mu + \epsilon^{\mu\nu\alpha\beta} \left[\frac{c}{6}\, \mu\, U_\nu\, \partial_\alpha (\mu\, U_\beta) + \frac{c}{2}\, \mu\, U_\nu\, \partial_\alpha A_\beta \right] \tag{10.52}$$

where μ is the chemical potential corresponding to the particle number and P is the pressure. Notice that T^μ_ν has the perfect fluid form. These Eqs. (10.51) and (10.52) were written down in [38] as a minimal way to incorporate anomalies. We now turn to the question of whether we can find an action which leads to these equations. We may expect such an action in terms of the formalism we have developed. Indeed such an action can be found, it is given by [39]

$$
S = \int d^4x \left[j^\mu (V_\mu + A_\mu) + \frac{c}{6} \epsilon^{\mu\nu\alpha\beta} \left(A_\mu V_\nu \partial_\alpha V_\beta + V_\mu A_\nu \partial_\alpha A_\beta \right) \right.
$$
$$
\left. - \mu \sqrt{-j^2} + P(\mu) \right] \tag{10.53}
$$

where $V_\mu = \partial_\mu \theta + \alpha \, \partial_\mu \beta$ and the flow velocity U^μ is related to V_μ by

$$
(V + A)_\mu = -\mu U_\mu \tag{10.54}
$$

It is not difficult to see why the action is of the form (10.53). The terms representing the anomaly must be independent of the metric, and hence it must be a differential four-form. The only one-forms available are the electromagnetic gauge potential $A = A_\mu \, dx^\mu$ and the velocity of the fluid for which we can use the Clebsch form, $V = V_\mu \, dx^\mu = d\theta + \alpha \, d\beta$. Thus we can take a linear combination of $A \, V \, dV$ and $V \, A \, dA$. The coefficients can be fixed by comparison with (10.51) and (10.52). This leads to the action (10.53). The equations which follow from this action have been analyzed in more detail in [39].

10.5.2   Anomalies in the Fluid Phase of the Standard Model

A more interesting scenario is where there are no gauge anomalies and we ask the question of how we can include the anomalies for the non-gauge symmetries. The most physical realization of this would be the standard model, so we will phrase our arguments in terms of it. We may regard the fluid we are talking about as the quark-gluon plasma phase for three flavors of quarks, say, u, d, s. In other words, we consider a phase with thermalized u, d, s quarks, so that they must be described by fluid variables while the heavier quarks are described by the field corresponding to each species. We will also neglect the quark masses so that we have the full flavor symmetry $U(3)_L \times U(3)_R$. Thus the group G to be used in (10.26) is

$$
G = SU(3)_c \times U(3)_L \times U(3)_R \tag{10.55}
$$

with individual flows corresponding to the charges. Here we want to focus on the flavor transport, as this is the sector with anomalies, so we will drop the color group $SU(3)_c$ from the equations to follow.

The flavor symmetry is not fully preserved even in the absence of masses; this is because of the anomalies. It may be useful at this point to recall the argument why we expect a term in the effective action which reproduces anomalies [40]. We set up a *gedanken* argument, where we consider all flavor symmetries to be gauged with their anomalies canceled by an extra set of fermions; the latter will not play any role in the dynamics except for the anomalies, so they are referred to as spectator fermions. The full theory is nonanomalous. The usual argument is that if, instead of the quarks, we consider the confined phase with mesons and baryons as the basic degrees of freedom, the theory will continue to remain nonanomalous. Even though the confined phase is obtained only at low energy, anomalies, because of their topological origin, are unaffected. Thus in the effective action for baryons and mesons, we should be able to find a term which reproduces the original anomalies, thereby ensuring cancellation with the spectator fermions. This is the Wess-Zumino term written in terms of the pseudoscalar meson fields. Clearly, we can expect a similar reasoning for the fluid phase where u, d, s are replaced by fluid variables. We must then have a term in the fluid action which can reproduce the anomalies so that the cancellation with spectator fermions still remains valid. How do we write this term? Since we have formulated fluid dynamics in terms of group-valued variables, the solution is almost trivial. We can simply use the usual Wess-Zumino term, but interpret the group-valued variables in it, not in terms of mesons, but as describing the fluid flow velocities for various flavor quantum numbers.

Adapting (10.26) to the case at hand with $U(3)_L \times U(3)_R$ symmetry, the action for fluid phase of the standard model is [36]

$$
\begin{aligned}
S = \int \Bigg[&-i j_3^\mu \, \mathrm{Tr}\left(t_3 \, g_L^{-1} D_\mu \, g_L\right) - i j_8^\mu \, \mathrm{Tr}\left(t_8 \, g_L^{-1} D_\mu \, g_L\right) \\
&- i k_3^\mu \, \mathrm{Tr}\left(t_3 \, g_R^{-1} D_\mu \, g_R\right) - i k_8^\mu \, \mathrm{Tr}\left(t_8 \, g_R^{-1} D_\mu \, g_R\right) \\
&- i j_0^\mu \, \mathrm{Tr}\left(g_L^{-1} D_\mu \, g_L\right) - i k_0^\mu \, \mathrm{Tr}\left(g_R^{-1} D_\mu \, g_R\right) \\
&- F(j_l \cdot j_{l'}, k_l \cdot k_{l'}, j_l \cdot k_{l'}) \Bigg] + S_{YM}(A) \\
&+ \Gamma_{wz}(A_L, A_R, g_L \, g_R^\dagger) - \Gamma_{wz}(A_L, A_R, \mathbb{1})
\end{aligned}
\tag{10.56}
$$

The three diagonal generators correspond to the t_3, t_8 and the identity for $U(3)_L$ and $U(3)_R$, with the corresponding currents j_3^μ, j_8^μ, j_0^μ and k_3^μ, k_8^μ, k_0^μ. The function F can in general depend on all invariants of the form $j_l \cdot j_{l'} = j_l^\mu j_{\mu l'}$, $k_l \cdot k_{l'}$, $j_l \cdot k_{l'}$, $l, l' = 0, 3, 8$.

The group elements $g_L \in U(3)_L$ and $g_R \in U(3)_R$ will describe the various flow velocities; their relation to the currents is seen upon eliminating the latter by the equations of motion. Further, $\Gamma_{wz}(A_L, A_R, g_L \, g_R^\dagger)$ is the standard Wess-Zumino term $\Gamma_{wz}(A_L, A_R, U)$ with U replaced by $g_L \, g_R^\dagger$. We have also subtracted $\Gamma_{wz}(A_L, A_R, \mathbb{1})$ which is necessary to bring the analysis to the so-called Bardeen form of the anomalies [41]. The Bardeen form is the one which not only preserves the vector gauge symmetries, but also gives a manifestly vector-gauge-invariant form to the remaining axial anomalies. This form is what is appropriate for the fluid phase. The explicit expression for $\Gamma_{wz}(A_L, A_R, g_L \, g_R^\dagger)$ is

$$
\Gamma_{\mathrm{wz}} = - \frac{iN}{240\pi^2} \int_D \mathrm{Tr}(dU\,U^{-1})^5
$$

$$
- \frac{iN}{48\pi^2} \int_{\mathcal{M}} \mathrm{Tr}[(A_L\,dA_L + dA_L\,A_L + A_L^3)\,dU\,U^{-1}]
$$

$$
- \frac{iN}{48\pi^2} \int_{\mathcal{M}} \mathrm{Tr}[(A_R\,dA_R + dA_R\,A_R + A_R^3)\,U^{-1}dU]
$$

$$
+ \frac{iN}{96\pi^2} \int_{\mathcal{M}} \mathrm{Tr}[A_L\,dU\,U^{-1}\,A_L\,dU\,U^{-1} - A_R\,U^{-1}dU\,A_R\,U^{-1}dU]
$$

$$
+ \frac{iN}{48\pi^2} \int_{\mathcal{M}} \mathrm{Tr}\Big[A_L(dU\,U^{-1})^3 + A_R(U^{-1}dU)^3
$$

$$
+ dA_L\,dU\,A_R\,U^{-1} - dA_R\,d(U^{-1})\,A_L\,U\Big]
$$

$$
+ \frac{iN}{48\pi^2} \int_{\mathcal{M}} \mathrm{Tr}[A_R\,U^{-1}\,A_L\,U(U^{-1}dU)^2 - A_L\,U\,A_R\,U^{-1}(dU\,U^{-1})^2]
$$

$$
- \frac{iN}{48\pi^2} \int_{\mathcal{M}} \mathrm{Tr}\Big[(dA_R\,A_R + A_R\,dA_R)\,U^{-1}\,A_L\,U
$$

$$
- (dA_L\,A_L + A_L\,dA_L)\,U\,A_R\,U^{-1}\Big]
$$

$$
- \frac{iN}{48\pi^2} \int_{\mathcal{M}} \mathrm{Tr}[A_L\,U\,A_R\,U^{-1}\,A_L\,dU\,U^{-1} + A_R\,U^{-1}\,A_L\,U\,A_R\,U^{-1}dU]
$$

$$
- \frac{iN}{48\pi^2} \int_{\mathcal{M}} \mathrm{Tr}\Big[A_R^3\,U^{-1}\,A_L\,U - A_L^3\,U\,A_R\,U^{-1}
$$

$$
+ \tfrac{1}{2}U\,A_R\,U^{-1}\,A_L\,U\,A_R\,U^{-1}\,A_L\Big] \tag{10.57}
$$

with $U = g_L\,g_R^\dagger$. (N is the number of colors, $= 3$ for us.) This is evidently a very complicated expression and we will need to pick out some pieces to highlight some physical effects. The most relevant of such effects is the chiral magnetic effect.

10.5.3   The Chiral Magnetic Effect

The chiral magnetic effect corresponds to the following. In the quark-gluon plasma, in the presence of a magnetic field, there is charge separation and a chiral induction which may be displayed as

$$
J_0 = \frac{e^2}{2\pi^2}\,\nabla\theta \cdot \boldsymbol{B}, \qquad J_i = -\frac{e^2}{2\pi^2}\,\dot{\theta}\,B_i \tag{10.58}
$$

Here θ is an axial $U(1)$ field, similar to the η'-meson. In the plasma, we can replace $\dot{\theta}$ by the difference of the chemical potentials μ_L and μ_R corresponding to the $U(1)_L$ and $U(1)_R$ subgroups of $U(3)_L \times U(3)_R$ as $\dot{\theta} \to \tfrac{1}{2}(\mu_L - \mu_R)$. In this case, we find

$$J_i = -\frac{e^2}{4\pi^2} \left(\mu_\text{L} - \mu_\text{R} \right) B_i \qquad (10.59)$$

We see that the chiral asymmetry of chemical potentials can lead to an electromagnetic current in the direction of the magnetic field [42]. In the experiment with colliding heavy nuclei which produces this fluid phase, if the collision is slightly off-center, the two nuclei constitute a current which produces, for a very short time, an intense magnetic field of the order of 10^{17} G. The resulting current can affect the charge distribution of particles, creating an asymmetry in the total charge of particles coming off in the direction of $\boldsymbol{B}$ versus $-\boldsymbol{B}$. Such an asymmetry is indeed experimentally observed; however, there are other effects to be taken into account which could possibly give an alternate explanation. So it is not entirely clear if the observed asymmetry can be attributed to the chiral magnetic effect. We may also note that the original calculation of the chiral magnetic effect is via Feynman diagrams, but the fluid action readily incorporates this effect due to the anomaly.

There are many other anomaly related effects, such as a possible pion asymmetry [43] or chiral vorticity effects. But the present discussion suffices to illustrate the main issues of principle; the reader is referred to the cited references for details. The full set of equations following from (10.56) are necessary to describe full hydrodynamic transport of flavor charges.

The main thrust of this chapter was to show that the co-adjoint orbit action introduced in Chap. 6, which is the quintessential realization of geometric quantization, can be used for fluid dynamics and can lead to many new insights.

Problems

10.1 Write the Wilson line for an $SU(2)$ gauge field as a path integral. (Although this is based on material from Chap. 6, it is included here as it serves as a prelude to constructing flow equations for charged fluids.)

10.2 The dynamics of a particle moving on Anti-de Sitter space in 4+1 dimensions ($= SO(4, 2)/SO(4, 1)$) can be described by the co-adjoint orbit action

$$\mathcal{S} = \int dt \left[-i\frac{mR}{2} \text{Tr}(\gamma_0 g^{-1}\dot{g}) + \frac{s_1}{2}\text{Tr}(\gamma_1\gamma_2\, g^{-1}\dot{g}) + \frac{s_2}{2}\text{Tr}(\gamma_3\gamma_5\, g^{-1}\dot{g}) \right]$$

where γ_μ, $\mu = 0, 1, 2, 3, 5$, are the usual Dirac matrices and s_1, s_2 label the two spins needed since $SO(4)$ is of rank 2. The group element may be parametrized as

$$g = \begin{pmatrix} \sqrt{z} & iX/\sqrt{z} \\ 0 & 1/\sqrt{z} \end{pmatrix} \Lambda, \qquad X = x^0 - \boldsymbol{\sigma} \cdot \boldsymbol{x}, \qquad \Lambda = e^{-[\gamma_\mu, \gamma_\nu]\Theta^{\mu\nu}}$$

Identify the analog of (10.46) for this case and formulate fluid dynamics for mass and spin flows on AdS$_5$.

Open Access This chapter is licensed under the terms of the Creative Commons Attribution 4.0 International License (http://creativecommons.org/licenses/by/4.0/), which permits use, sharing, adaptation, distribution and reproduction in any medium or format, as long as you give appropriate credit to the original author(s) and the source, provide a link to the Creative Commons license and indicate if changes were made.

The images or other third party material in this chapter are included in the chapter's Creative Commons license, unless indicated otherwise in a credit line to the material. If material is not included in the chapter's Creative Commons license and your intended use is not permitted by statutory regulation or exceeds the permitted use, you will need to obtain permission directly from the copyright holder.

Chapter 11
Quantization Rules

In the usual textbook discussions of quantum mechanics, one usually starts with a quantization rule which assigns an operator for any classical function on the phase space. The simplest example of this is the correspondence

$$x \to \hat{x}, \qquad p \to \hat{p} = -i\hbar \frac{\partial}{\partial x} \tag{11.1}$$

For arbitrary functions of x and p, the simple use of this rule can lead to operator ordering issues. Thus the function xp could be viewed as $\hat{x}\hat{p}$, $\hat{p}\hat{x}$ or $\frac{1}{2}(\hat{x}\hat{p} + \hat{p}\hat{x})$. The first two versions are not hermitian. The operator $\frac{1}{2}(\hat{x}\hat{p} + \hat{p}\hat{x})$ is the Weyl-ordered version of xp. More generally for the Weyl-ordered version of a function $f(x, p)$, one can use

$$\hat{f} = \int \frac{dx\,dp}{2\pi} \frac{du\,dv}{2\pi} e^{iu(\hat{x}-x)+iv(\hat{p}-p)} f(x, p) \tag{11.2}$$

Effectively, here one is taking the Fourier transform of $f(x, p)$ and then transforming back with operators $\hat{x}$, $\hat{p}$ in place of x, p. For polynomials of x and p, this leads to the symmetrized products of the operators which are also hermitian.

There is also an inversion formula due to Wigner for (11.2). Taking the matrix element of the operator $\hat{f}$ in an x-diagonal basis of states, the inversion formula is[1]

$$f(x, p) = \int dz\, e^{-ipz/\hbar} \langle x + \tfrac{1}{2}z | \hat{f} | x - \tfrac{1}{2}z \rangle \tag{11.3}$$

In the context of geometric quantization, there is a very elegant procedure for the function-to-operator correspondence known as the Berezin-Toeplitz quantiza-

---

[1] We kept $\hbar$ in (11.1) and (11.3), so the reader can see how an $\hbar$-expansion can be worked out.

© The Author(s) 2024
V. P. Nair, *Geometric Quantization and Applications to Fields and Fluids*,
SpringerBriefs in Physics, https://doi.org/10.1007/978-3-031-65801-3_11

tion [44]. When the phase space is a complex Kähler manifold (with complex coordinates $z_\alpha, \bar{z}_\alpha$), we can consider the geometric quantization of a suitable Ω in the holomorphic polarization. Let $\{\psi_i\}$ form a basis of wave functions. Then the matrix elements of an operator $\hat{A}$ are given in this basis as

$$A_{ij} = \int d\mu \, \psi_i^* \, A(z, \bar{z}) \, \psi_j \tag{11.4}$$

This is the Berezin-Toeplitz quantization rule. Here $A(z, \bar{z})$ is a function on the phase space M, known as the contravariant symbol of the operator $\hat{A}$.

There are a couple of questions which arise naturally in this context. The first is about how operator products work out in relation to the classical theory [44, 45]. The second question is about the existence of an inverse relation to (11.4) where we can construct $A(z, \bar{z})$ given the matrix elements A_{ij}. We will consider these questions in turn.

The matrix elements of the products $\hat{A}\hat{B}$ of two operators $\hat{A}$ and $\hat{B}$ are given by

$$(\hat{A}\hat{B})_{ij} = \sum_k A_{ik} B_{kj} = \int d\mu d\mu' \, \psi_i^*(z) A(z, \bar{z}) \, K(z, z') \, B(z', \bar{z}') \psi_j(z')$$

$$K(z, z') = \sum_k \psi_k(z) \psi_k^*(z') \tag{11.5}$$

The reduction of the kernel $K(z, z')$ is what is needed to simplify this expression.

Consider, as an example, the case of the complex plane $\mathbb{C}$ as the phase space with the symplectic form $\Omega = (i/\kappa) dz \wedge d\bar{z}$. As seen in Chap. 6, a basis for the wave functions in this case is given by

$$\psi_n = e^{-\frac{1}{2}(z\bar{z}/\kappa)} \frac{z^n}{\kappa^{\frac{n}{2}} \sqrt{n!}} \tag{11.6}$$

We can then write (11.5) as

$$(\hat{A}\hat{B})_{nm} = \frac{\kappa^{-\frac{n+m}{2}}}{\sqrt{n!m!}} \int d\mu d\mu' \, \bar{z}^n A(z, \bar{z}) e^{(z\bar{z}' - z\bar{z} - z'\bar{z}')/\kappa} B(z', \bar{z}') z'^m$$

$$= \frac{\kappa^{-\frac{n+m}{2}}}{\sqrt{n!m!}} \int d\mu d\mu' \, \bar{z}^n A(z, \bar{z}) e^{-(z\bar{z} + z'\bar{z}')/\kappa} B(z' + z, \bar{z}')(z' + z)^m$$

$$= \frac{\kappa^{-\frac{n+m}{2}}}{\sqrt{n!m!}} \int d\mu d\mu' \, \bar{z}^n A(z, \bar{z}) e^{-(z\bar{z} + z'\bar{z}')/\kappa} e^{z'\partial_z} [B(z, \bar{z}') z^m] \tag{11.7}$$

For the integral over $z', \bar{z}'$, taking B to be power-expandable in $\bar{z}$, we can write

$$\int d\mu'\, e^{-z'\bar{z}/\kappa}\, e^{z'\partial_z}\left[B(z,\bar{z}')z^m\right] = \int e^{-z'\bar{z}'\kappa}\, e^{z'\partial_z}\sum_s \frac{1}{s!}\bar{z}'^s \partial_{\bar{w}}^s B(z,\bar{w})z^m\bigg|_{\bar{w}=0}$$

$$= \sum_s \frac{\kappa^s}{s!}(\partial_z\partial_{\bar{w}})^s\left(B(z,\bar{w})z^m\right)\bigg|_{\bar{w}=0}$$

$$= e^{\kappa\partial_z\partial_{\bar{w}}}\left(B(z,\bar{w})z^m\right)\bigg|_{\bar{w}=0} \tag{11.8}$$

In combining this with the rest of the terms in (11.7), we can use

$$e^{-z\bar{z}/\kappa}e^{\kappa\partial_z\partial_{\bar{w}}}B(z,\bar{w}) = e^{\kappa\partial_z\partial_{\bar{w}}}e^{-z\bar{z}/\kappa}e^{\bar{z}\partial_{\bar{w}}}B(z,\bar{w})$$

$$= e^{\kappa\partial_z\partial_{\bar{w}}}e^{-z\bar{z}/\kappa}B(z,\bar{z}+\bar{w}) \tag{11.9}$$

We can now substitute this into (11.7) and and do integrations by parts to transfer the ∂_z derivatives to act on $A(z,\bar{z})$. Notice that the factor $\bar{z}^n$ is not affected by this. This is the advantage of the holomorphic polarization. The result is

$$(\hat{A}\hat{B})_{nm} = \frac{\kappa^{-\frac{n+m}{2}}}{\sqrt{n!m!}}\int d\mu\, \bar{z}^n A(z,\bar{z})e^{\kappa\partial_z\partial_{\bar{w}}}e^{-z\bar{z}/\kappa}B(z,\bar{z}+\bar{w})z^m\bigg|_{\bar{w}=0}$$

$$= \frac{\kappa^{-\frac{n+m}{2}}}{\sqrt{n!m!}}\int d\mu\, \bar{z}^n e^{-z\bar{z}/\kappa}\left[\sum_s(-\kappa)^s(\partial_z^s A\, \partial_{\bar{z}}^s B)\right]z^m$$

$$= \int d\mu\, \psi_n^*\,(A*B)\,\psi_m \tag{11.10}$$

This shows that the symbol corresponding to the product of the operators is given by the "star-product" of the symbols for the operators, which is defined by

$$A*B = \sum_s(-\kappa)^s(\partial_z^s A\, \partial_{\bar{z}}^s B)$$

$$= AB - \kappa\,\partial_z A\,\partial_{\bar{z}}B + \mathcal{O}(\kappa^2) \tag{11.11}$$

The star-product is given by the ordinary commuting product of the functions plus terms involving derivatives of the functions. The $*$-commutator takes the form

$$A*B - B*A = \kappa\,(\partial_{\bar{z}}A\partial_z B - \partial_z A\partial_{\bar{z}}B) + \mathcal{O}(\kappa^2)$$

$$= \mathrm{i}\{A,B\} + \mathcal{O}(\kappa^2) \tag{11.12}$$

where we have used the definition of the Poisson bracket for the given Ω as

$$\{A,B\} = \mathrm{i}\kappa\,(\partial_{\bar{z}}A\partial_z B - \partial_z A\partial_{\bar{z}}B) + \mathcal{O}(\kappa^2) \tag{11.13}$$

In other words, the symbol corresponding to the commutator is i times the Poisson bracket, which is the usual and expected correspondence. (Small values of κ correspond to the semiclassical limit.)

Similar reasoning will apply to other Kähler manifolds as well [45]. Thus for a coset manifold of the form G/H, the wave functions are given by

$$\Psi_i^{(r)} = \sqrt{N} \, \langle r, i | \hat{g} | r, w \rangle = \sqrt{N} \, \mathcal{D}_{i,w}^{(r)}(g) \tag{11.14}$$

where N is the dimension of the representation r and the state $|r, w\rangle$ is a highest weight state chosen by the polarization condition. The matrix elements of an operator take the form as in (11.4), and the relevant kernel for the star-product is

$$K(g, g') = N \langle r, w | \hat{g}'^{\dagger} \hat{g} | r, w \rangle \tag{11.15}$$

The reduction of this needed to write the star-product in a form analogous to (11.11) is somewhat more involved. The simplest way is to use a complete set of eigenfunctions of the Laplace operator on the manifold [46]. For the case of $SU(2)/U(1)$ with $\Omega = -\mathrm{i}(n/2)\mathrm{Tr}(\sigma_3 g^{-1}dg\, g^{-1}dg)$, for example, the star-product takes the form

$$A(z, \bar{z}) * B(z, \bar{z}) = A(z, \bar{z}) \, B(z, \bar{z}) + \sum_{s=1}^{\infty} (-1)^s c_s \left(R_+^s A(z, \bar{z}) \right) \left(R_-^s \, B(z', \bar{z}') \right)$$

$$= A\, B + \frac{1}{(n+2)} (R_+ A)\, (R_- B)$$

$$+ \frac{1}{2(n+2)(n+3)} (R_+^2 A)\, (R_-^2 B) + \cdots \tag{11.16}$$

where $R_\pm$ are the right translation operators on g, introduced in Chap. 6. The coefficients c_s in this expansion can be recursively calculated by successive application of the completeness relation for the eigenfunctions of the Laplacian. We see from this equation that we obtain the Poisson bracket as the $*$-commutator in large n limit, which is the semiclassical limit in this case.

Consider now the question of constructing the function (or symbol) from the matrix elements of the operator. We will use a coset manifold of the G/H-type (with G and H compact) for this. The result for the case of the plane (or higher dimensional flat spaces) can be obtained by a suitable large radius limit. Given the symmetry G, we can consider a tensor operator which transforms according to the irreducible representation r' of G, i.e.,

$$\hat{g}\, \hat{F}_\alpha\, \hat{g}^{\dagger} = \sum_{\beta} \hat{F}_\beta \, \langle r', \beta | \hat{g} | r', \alpha \rangle \tag{11.17}$$

For the matrix elements of such an operator, we have the Wigner-Eckart theorem,

$$(\hat{F}_\alpha)_{ij} = \langle r, i | r', \alpha; r, j \rangle \, \langle\!\langle \hat{F} \rangle\!\rangle \tag{11.18}$$

where $\langle r, i | r', \alpha; r, j \rangle$ is the Clebsch-Gordan coefficient relating the product of the representations r' and r to r and $\langle\!\langle F \rangle\!\rangle$ is the reduced matrix element defined as

$$\langle\!\langle \hat{F} \rangle\!\rangle = \frac{1}{N} \sum_{i,j,\beta} \langle r', \beta; r, j | r, i \rangle \, (\hat{F}_\beta)_{ij} \tag{11.19}$$

In terms of the representation matrices for g, we have the relation

$$\int d\mu(g) \mathcal{D}^{(r)}_{i,i'}(g^\dagger) \, \mathcal{D}^{(r)}_{j',j}(g) \mathcal{D}^{(r')}_{\beta,\alpha}(g) = \frac{1}{N} \langle r', \beta; r, j' | r, i' \rangle \langle r, i | r', \alpha; r, j \rangle . \tag{11.20}$$

Setting $\beta = 0$, $i' = j' = w$ in this equation, we can write

$$\langle\!\langle \hat{F} \rangle\!\rangle \, \langle r, i | r', \alpha; r, j \rangle \, \langle r', 0; r, w | r, w \rangle$$

$$= N \int d\mu(g) \, \mathcal{D}^{(r)*}_{w,i}(g^\dagger) \, \mathcal{D}^{(r)}_{w,j}(g) \, \langle\!\langle \hat{F} \rangle\!\rangle \mathcal{D}^{(r')}_{0,\alpha}(g)$$

$$= N \int d\mu(g) \, \mathcal{D}^{(r)*}_{i,w}(g^T) \mathcal{D}^{(r)}_{j,w}(g^T) \, \langle\!\langle \hat{F} \rangle\!\rangle \mathcal{D}^{(r')}_{\alpha,0}(g^T)$$

$$= \int d\mu(g) \, \Psi^{(r)*}_i \left[\langle\!\langle \hat{F} \rangle\!\rangle \mathcal{D}^{(r')}_{\alpha,0}(g) \right] \Psi^{(r)}_j \tag{11.21}$$

where, in the last step, we changed the variables of integration as $g \to g^T$, and used the definition of wave functions in (11.14). We can now define a function $F_\alpha(g)$ by

$$F_\alpha(g) \, \langle r', 0; r, w | r, w \rangle = \mathcal{D}^{(r')}_{\alpha,0}(g) \langle\!\langle \hat{F} \rangle\!\rangle . \tag{11.22}$$

Equation (11.21) then takes the form

$$\int d\mu(g) \, \Psi^{(r)*}_i \, F_\alpha(g) \, \Psi^{(r)}_j = \langle r, i | r', \alpha; r, j \rangle \, \langle\!\langle \hat{F} \rangle\!\rangle$$

$$= (\hat{F}_\alpha)_{ij} . \tag{11.23}$$

We see that the matrix element of $\hat{F}_\alpha$ is reproduced in terms of a function $F_\alpha(g)$, in accordance with the Berezin-Toeplitz formula (11.4). Equation (11.22), along with (11.18), is the required inverse relation to (11.4), constructing the contravariant symbol in terms of the matrix elements of the operator. For a general operator, one can first write it as a linear combination of tensor operators and the result extends by linearity; i.e.,

$$\hat{F} = \sum_{r',\alpha} C_{r',\alpha} \, \hat{F}^{(r')}_\alpha \implies F = \sum_{r',\alpha} C_{r',\alpha} \, F^{(r')}_\alpha \tag{11.24}$$

The existence of a function in terms of which one can write the matrix elements of an operator is essentially the diagonal coherent state representation, originally proved by Sudarshan for coherent states in flat space [47]. The arguments given above were taken from [48].

For completeness, we may note that there is another notion of a function associated with an operator known as the covariant symbol. It is defined by

$$(A) = \frac{1}{N} \sum_{ij} \Psi_i A_{ij} \Psi_j^* = \sum_{ij} \mathcal{D}_{i,w}^{(r)}(g) A_{ij} \mathcal{D}_{j,w}^{(r)*}(g) \qquad (11.25)$$

This is different from the contravariant symbol in the sense that if we start from $A(z, \bar{z})$, construct A_{ij} according to (11.4) and then construct (A) as in (11.25), then $(A) \neq A(z, \bar{z})$ in general. The classical limits will be the same; for example, for the case of $SU(2)/U(1)$,

$$(A) = A(z, \bar{z}) + \text{terms of order } \frac{1}{n} \qquad (11.26)$$

A star-product can be defined for the covariant symbols as well. The $*$-commutator will reproduce the Poisson bracket, differing from the $*$-commutator for the contravariant symbols only at order κ^2 or at $1/n^2$.

There is a close connection between Berezin-Toeplitz quantization, covariant and contravariant symbols, Landau levels for the quantum Hall effect and fuzzy spaces. It is difficult to survey this vast field here, the articles [48, 49] and references therein can facilitate an entrée into the large body of literature.

Open Access This chapter is licensed under the terms of the Creative Commons Attribution 4.0 International License (http://creativecommons.org/licenses/by/4.0/), which permits use, sharing, adaptation, distribution and reproduction in any medium or format, as long as you give appropriate credit to the original author(s) and the source, provide a link to the Creative Commons license and indicate if changes were made.

The images or other third party material in this chapter are included in the chapter's Creative Commons license, unless indicated otherwise in a credit line to the material. If material is not included in the chapter's Creative Commons license and your intended use is not permitted by statutory regulation or exceeds the permitted use, you will need to obtain permission directly from the copyright holder.

Chapter 12
A Comment on the Metaplectic Correction

In Sect. 7.1, we considered the quantization of the symplectic form $\Omega = i \, dz \wedge d\bar{z}$, obtaining the standard coherent states. (In this chapter, we are setting $\kappa = 1$ for convenience.) The quantum operator corresponding to $\bar{z}z$ was also identified as $z\partial_z + \frac{1}{2}$, where the extra term $\frac{1}{2}$ is due to the metaplectic correction. We now consider a set of symplectic transformations which can elucidate the meaning and importance of the metaplectic structure.

The symplectic form $\Omega = i \, dz \wedge d\bar{z}$ is invariant under the infinitesimal transformations

$$z \to z' = z + i \, A \, z + B \, \bar{z}, \qquad \bar{z} \to \bar{z}' = \bar{z} - i \, A \, \bar{z} + \bar{B} \, z \qquad (12.1)$$

where A is real. The finite version of these transformations form the $Sp(1, \mathbb{R})$ group. We can also introduce real variables (p, q) by

$$z = \frac{1}{\sqrt{2}} \, (p + iq), \qquad \bar{z} = \frac{1}{\sqrt{2}} \, (p - iq) \qquad (12.2)$$

for which $\Omega = dp \wedge dq$. This choice of coordinates would be convenient for choosing real polarizations such as wave functions which only depend on q. The transformations (12.1) do not preserve holomorphicity and these are what help to connect different polarizations. For example, consider for simplicity the case of $A = 0$. Then differentiating $f(z', \bar{z}')$, we find

$$\partial_{\bar{z}} \approx \partial_{\bar{z}'} + B \, \partial_{z'}, \qquad \partial_{\bar{z}'} \approx \partial_{\bar{z}} - B \, \partial_z \qquad (12.3)$$

We see that the holomorphic polarization in terms of z, $\bar{z}$ is not the same as the holomorphic polarization in terms of z', $\bar{z}'$. This shows that the transformations (12.1) help implement infinitesimal changes of polarization, which can lead to real polarizations or polarizations which are not holomorphic. Classically, we have a closed Poisson bracket algebra for the generators of the transformations,

© The Author(s) 2024
V. P. Nair, *Geometric Quantization and Applications to Fields and Fluids*,
SpringerBriefs in Physics, https://doi.org/10.1007/978-3-031-65801-3_12

$$\{f_A, f_B\} = -\mathrm{i}\,f_B, \qquad \{f_A, f_{\bar{B}}\} = \mathrm{i}\,f_{\bar{B}}$$
$$\{f_{\bar{B}}, f_B\} = -2\mathrm{i}\,f_A \tag{12.4}$$

for $f_A = \frac{1}{2}z\bar{z}$, $f_B = -(\mathrm{i}/2)z^2$, $f_{\bar{B}} = (\mathrm{i}/2)\bar{z}^2$. This is the algebra of $Sp(1, \mathbb{R})$.

In going to the quantum theory, since we need to have the facility of changing polarizations, the unitary implementation of (12.1) *is important.* The operators corresponding to $\bar{z}, z$ are the annihilation and creation operators $a, a^\dagger$, respectively, with $[a, a^\dagger] = 1$. In the holomorphic polarization, this corresponds to the identification

$$z \to a^\dagger = z \qquad \bar{z} \to a = \frac{\partial}{\partial z}, \tag{12.5}$$

so that the quantum version of $z\bar{z}$ without the metaplectic correction is $a^\dagger a$.

However, the quantum version of $f_B = -(\mathrm{i}/2)z^2$, $f_{\bar{B}} = (\mathrm{i}/2)\bar{z}^2$ are unambiguously given by the prequantum operators as

$$\hat{f}_B = -\frac{\mathrm{i}}{2}a^{\dagger 2}, \qquad \hat{f}_{\bar{B}} = \frac{\mathrm{i}}{2}a^2 \tag{12.6}$$

Their commutator is given by

$$[\hat{f}_{\bar{B}}, \hat{f}_B] = \left(a^\dagger a + \frac{1}{2}\right) = 2\left[\tfrac{1}{2}(a^\dagger a + \tfrac{1}{2})\right]$$
$$\equiv 2\,\hat{f}_A \tag{12.7}$$

We see that the closure of the algebra and the quantum implementation of (12.6) requires us to identify $\hat{f}_A = \frac{1}{2}[a^\dagger a + \frac{1}{2}]$ as the quantum generator of the A-type transformations. In other words, we should have $z\bar{z} \to (a^\dagger a + \frac{1}{2})$. *The essence of the metaplectic correction is thus the quantum realization of the $Sp(1, \mathbb{R})$.*

Notice that while the quantum operator corresponding to $\bar{z}z$ is identified as $a^\dagger a + \frac{1}{2}$, there is no statement about whether one should use this operator for a Hamiltonian. We bring up this point because, sometimes in the literature, one finds the statement that the half-form quantization is needed as it leads to the "correct" quantization which should have the zero-point energy if one applies this to the harmonic oscillator (for which the classical Hamiltonian is $\bar{z}z$). This statement certainly needs some clarification. The classical Hamiltonian for the oscillator is $\bar{z}z + C$ for any constant C, so the question of zero-point energy is completely different. To sharpen this point, consider the free relativistic scalar field which can be considered as a collection of harmonic oscillators. In fact, for scalar fields in a cubical box of volume V with periodic boundary conditions, we can use the mode expansion

$$\phi(x) = \sum_k Z_k\,u_k(x) + \bar{Z}_k\,u_k^*(x)$$

$$\dot{\phi}(x) = \pi(x) = \sum_k (-i\omega_k) \left(Z_k \, u_k(x) - \bar{Z}_k \, u_k^*(x) \right)$$

$$u_k(x) = \frac{1}{\sqrt{2\omega_k V}} \, e^{-ik \cdot x}, \qquad \omega_k = \sqrt{k^2 + m^2} \qquad (12.8)$$

The phase space degrees of freedom correspond to $(Z_k, \bar{Z}_k)$. From the action of a free scalar field, we find the symplectic structure and classical Hamiltonian as

$$\Omega = i \prod_k dZ_k \wedge d\bar{Z}_k, \qquad H = \sum_k \omega_k \, \bar{Z}_k Z_k + C \qquad (12.9)$$

Classically, this is indeed a collection of harmonic oscillators. However, in quantizing this, keeping any nonzero value for the zero-point energy is the wrong thing to do. For this problem, we want to obtain a unitary realization of the Poincaré group. One of the commutation rules for this group is

$$[P_i, K_j] = i \, \delta_{ij} \, H \qquad (12.10)$$

where P_i is the momentum operator and K_j is the Lorentz boost generator. The Lorentz invariance of the vacuum (in the limit of $V \to \infty$) requires $K_j |0\rangle = 0$. As a result, we must have $\langle 0| \, H \, |0\rangle = 0$ (upon taking the expectation value of (12.10)), showing that the quantization we need should have no zero-point energy. The correct thing to do is thus to define a renormalized form of the quantum Hamiltonian which has no zero point energy.

Notice that the generators of the symplectic transformations discussed earlier do have the extra metaplectic correction, but the choice of the Hamiltonian (and how it should represented as an operator) is determined by imposing desirable symmetries, the Lorentz invariance of the vacuum in the present context. More explicitly, the relevant algebraic relations for the symplectic transformations are

$$[a_k a_l, a_r^\dagger a_s^\dagger] = \delta_{kr} a_s^\dagger a_l + \delta_{lr} a_s^\dagger a_k + \delta_{ks} a_r^\dagger a_l + \delta_{ls} a_r^\dagger a_k + (\delta_{ks}\delta_{lr} + \delta_{kr}\delta_{ls}) \quad (12.11)$$

One does realize this algebra unitarily on the Fock space of the theory. (The finite transformations corresponding (12.11) are also what are used to generate squeezed states in quantum optics.) The Hamiltonian however is one of the generators of the Poincaré algebra, given by $H = \sum_k \omega_k \, a_k^\dagger a_k$ (with no term corresponding to the zero-point energy) and $P_i = \sum_k k_i \, a_k^\dagger a_k$. For some recent work in relating the metaplectic correction to coherent state path integrals, see [50].

Open Access This chapter is licensed under the terms of the Creative Commons Attribution 4.0 International License (http://creativecommons.org/licenses/by/4.0/), which permits use, sharing, adaptation, distribution and reproduction in any medium or format, as long as you give appropriate credit to the original author(s) and the source, provide a link to the Creative Commons license and indicate if changes were made.

The images or other third party material in this chapter are included in the chapter's Creative Commons license, unless indicated otherwise in a credit line to the material. If material is not included in the chapter's Creative Commons license and your intended use is not permitted by statutory regulation or exceeds the permitted use, you will need to obtain permission directly from the copyright holder.

Solutions to Problems

Problem 2.1 Derive Eq. (2.18) from the definition of Poisson brackets.

Solution:

$$
\begin{aligned}
i_\xi i_\eta i_\rho(\mathrm{d}\Omega) &= \xi^\nu \eta^\mu \rho^\alpha (\partial_\alpha \Omega_{\mu\nu} + \partial_\mu \Omega_{\nu\alpha} + \partial_\nu \Omega_{\alpha\mu}) \\
&= \xi^\nu \rho^\alpha \left[\partial_\alpha(\eta^\mu \Omega_{\mu\nu}) - (\partial_\alpha \eta^\mu)\Omega_{\mu\nu} \right] + \text{cyclicperm.} \\
&= -\xi^\nu \partial_\nu(\rho^\alpha \partial_\alpha g) - \eta^\mu \partial_\mu(\xi^\nu \partial_\nu h) - \rho^\alpha \partial_\alpha(\eta^\mu \partial_\mu f) \\
&= \{f, \{g, h\}\} + \{g, \{h, f\}\} + \{h, \{f, g\}\}
\end{aligned}
$$

Problem 5.1 For a particle moving on a circle with coordinate θ, $d\theta/(2\pi)$ is an element of $\mathcal{H}^1(M)$. Consider the action

$$
S = \int dt \left[\tfrac{1}{2}\dot\theta^2 + \frac{\alpha}{2\pi}\dot\theta \right]
$$

Obtain the energy eigenvalues to show how they depend on the vacuum angle α.

Solution: The solution is straightforward, the eigenvalues are labeled by an integer n, with $E_n = \frac{1}{2}(n - (\alpha/2\pi))^2$.

Problem 6.1 Find the spin connection and curvature and its integral for S^2.

Solution: With no torsion, the spin connection $w^a{}_b$ is defined by $de^a + w^a{}_b \, e^b = 0$
From (6.15), we get

$$
w^1{}_2 = 2x \, e^2 - 2y \, e^2, \qquad w^2{}_1 = -w^1{}_2
$$

$$
R^1{}_2 = (\mathrm{d}w + w\,w)^1{}_2 = 4\, e^1 \, e^2, \qquad \int R^1{}_2 = 4\pi
$$

Problem 6.2 Carry out the infinitesimal transformations generated by the vector fields $\xi_\pm$, ξ_3 given in (6.27) and show that they are isometries of S^2.

© The Editor(s) (if applicable) and The Author(s) 2024
V. P. Nair, *Geometric Quantization and Applications to Fields and Fluids*,
SpringerBriefs in Physics, https://doi.org/10.1007/978-3-031-65801-3

Solution: The metric is given by $ds^2 = dz d\bar{z}/(1 + z\bar{z})^2$. The vector field ξ_+ corresponds to $z \to z + i\epsilon z^2$, $\bar{z} \to \bar{z} + i\epsilon$. The change in the metric is

$$ds'^2 = \frac{d(z + i\epsilon z^2) d(\bar{z} + i\epsilon)}{(1 + (z + i\epsilon z^2)(\bar{z} + i\epsilon))^2} \approx ds^2 + 2i\epsilon \frac{dz d\bar{z}}{(1 + z\bar{z})^2} - 2i\epsilon z(1 + z\bar{z}) \frac{dz d\bar{z}}{(1 + z\bar{z})^3}$$
$$\approx ds^2$$

Thus ξ_+ is an isometry; so is ξ_- since it is the conjugate and ds^2 is real. Further $[\xi_+, \xi_-] = 2i\xi_3$, so ξ_3 is also an isometry.

Problem 6.3 Identify the generators of left translations of g in Ω of (6.47).

Solution: With V as the generator of left translations of the form $Vg = \theta g$,

$$i_V \Omega = -i\frac{n}{2} \text{Tr} \left[\sigma_3 (g^{-1}\theta g g^{-1} dg - g^{-1} dg \, g^{-1} \theta g] = -i\frac{n}{2} d\text{Tr}(\sigma_3 g^{-1}\theta g)$$
$$= -d(Q_a \theta^a), \qquad Q_a = \frac{n}{2} \text{Tr}(\sigma_3 g^{-1} t_a g)$$

Problem 6.4 Consider the geometric quantization of the hyperboloidal space $SL(2, \mathbb{R})/U(1)$. The canonical two-form is given by

$$\Omega = 2i\lambda \frac{dz \wedge d\bar{z}}{(1 - z\bar{z})^2}$$

This applies to the region $z\bar{z} \leq 1$. Show that $V_+ = iz^2\partial_z - i\partial_{\bar{z}}$, $V_- = -i\bar{z}^2\partial_{\bar{z}} + i\partial_z$, $V_3 = iz\partial_z - i\bar{z}\partial_{\bar{z}}$ are Hamiltonian vector fields. Identify the nature of the wave functions, the inner product in the holomorphic polarization and the operators corresponding to $V_\pm$, V_3.

Solution: The wave functions are of the form $\psi = \mathcal{N}e^{-K/2} f(z)$, where the Kähler potential is $K = -2\lambda \log(1 - z\bar{z})$, with the inner product

$$\langle 1 | 2 \rangle = \frac{1}{2\pi i} \int d\bar{z} \wedge dz \frac{f_1^* f_2}{(1 - z\bar{z})^{2 - 2\lambda}}$$

The operators corresponding to $V_\pm$, V_3 are given by

$$J_+ f = (z^2\partial_z + 2\lambda z) f, \qquad J_- f = \partial_z f, \qquad J_3 f = (z\partial_z + \lambda) f$$

This should give the discrete series of representations of $SL(2, \mathbb{R})$ bounded below.

Problem 7.1 Calculate $\Omega(A^g) - \Omega(A)$ for finite transformations, i.e., obtain (7.37).

Solution: With δ denoting the exterior derivative along the gauge directions, $\delta A^g = g[\delta A + Dv] g^{-1}$, with $Dv = dv + Av + vA$, $v = g^{-1}\delta g$, and

$$\Omega(A^g) - \Omega(A) = \frac{k}{4\pi} \int \text{Tr}(\delta A Dv + Dv \delta A + Dv Dv)$$

$$\int \delta A D v = -\int (D\delta A)v = \int \delta F v, \qquad \int D v D v = \int v D^2 v = \int v(Fv - vF)$$

It should be kept in mind that δ and d anticommute. This gives

$$\Omega(A^g) - \Omega(A) = \frac{k}{4\pi} \int \mathrm{Tr}[(2\delta Fv) + vFv - v^2 F] = \delta \frac{k}{2\pi} \int \mathrm{Tr}(vF)$$

Problem 7.2 Derive the Polyakov-Wiegmann identity given in (7.54).

Solution: Straightforward expansion using $(Kh)^{-1}\mathrm{d}(Kh) = h^{-1}K^{-1}\mathrm{d}K\,h + h^{-1}\mathrm{d}h$ will give the result.

Problem 7.3 The WZW action can be quantized as a 1+1 dimensional field theory in its own right. In lightcone coordinates $u = (t - x)/\sqrt{2}$, $v = (t + x)/\sqrt{2}$, the action is

$$\mathcal{S} = -\frac{k}{4\pi} \int \mathrm{Tr}(\partial_u gg^{-1} \partial_v gg^{-1}) + \Gamma_{\mathrm{WZ}}$$

Identify the canonical two-form. Show that left translations of g, i.e., $g \to (1 + (-it_a\theta^a))g$ are generated by $J_v^a = (k/4\pi)(\partial_v gg^{-1})^a$. Obtain also the commutation rules

$$[J_v(\theta), J_v(\varphi)] = iJ_v(\theta \times \varphi) - i\frac{k}{4\pi} \int \partial_v \theta^a \varphi^a$$

Solution: From the general formula for the canonical one-form $\mathcal{A}$ given in (3.7), the canonical two-form should be

$$\Omega = \frac{k}{4\pi} \int \mathrm{d}v\, \mathrm{Tr}\left[\xi\partial_v\xi + 2\xi^2\partial_v gg^{-1}\right], \qquad \xi = \delta gg^{-1}$$

It is easy to check that J_v is the generator of left translations and the Poisson bracket then follows from the general formula (2.11).

Problem 8.1 Calculate $\nu[A]$ for the instanton in (8.13), (8.14).

Solution: For the given A, $\rho \to 0$ as $x_4 \to -\infty$, $\rho \to f(r)$ as $x_4 \to \infty$. Thus $A \to 0$ as $x_4 \to -\infty$ and $A \to -\mathrm{d}UU^{-1}$ as $x_4 \to \infty$, where

$$U = \phi^0 + i\sigma_i\phi^i, \qquad \phi^0 = \cos f, \quad \phi^i = \frac{x^i}{r}\sin f, \quad f(r) = \frac{\pi r}{\sqrt{r^2 + \alpha^2}}$$

The instanton number is simplified as

$$\nu[A] = -\frac{1}{8\pi^2} \int \mathrm{Tr} \left(A\mathrm{d}A + \frac{2}{3}A^3 \right) \Big|_{x^4=-\infty}^{|x^4=\infty} = -\frac{1}{24\pi^2} \int \mathrm{Tr}(\mathrm{d}UU^{-1})^3$$
$$= Q[U]$$

We also find $\mathrm{d}UU^{-1} = -i\sigma_i \left[(\phi^i \mathrm{d}\phi^0 - \mathrm{d}\phi^i \phi^0) - \epsilon_{ijk} \mathrm{d}\phi^j \phi^k \right]$ so that

$$\mathrm{Tr}(\mathrm{d}UU^{-1})^3 = \mathrm{Tr} \left[\mathrm{d}UU^{-1}\mathrm{d}(\mathrm{d}UU^{-1}) \right] = 6(\sin^2 f) \, \mathrm{d}f \, \epsilon_{ijk} \hat{x}^i \mathrm{d}\hat{x}^j \mathrm{d}\hat{x}^k$$
$$Q[U] = -\frac{1}{2\pi^2} \int (\sin^2 f) \mathrm{d}f \, \sin\theta \mathrm{d}\theta \mathrm{d}\varphi = \frac{1}{\pi} [f(0) - f(\infty)] = -1$$

Problem 9.1 Obtain the Landau levels for electrons in a uniform Abelian magnetic field on $\mathbb{CP}^2$ and show that the lowest level wave functions agree with (6.70).

Solution: Since $\mathbb{CP}^2 = SU(3)/U(2)$, the curvatures take values in the Lie algebra of $U(2) \sim SU(2) \times U(1)$ and are constant in the tangent frame basis. A uniform Abelian magnetic field can be taken to be proportional to the $U(1)$ curvature, so the one-particle wave functions obey

$$R_8 \, \psi = -\frac{n}{\sqrt{3}} \, \psi, \qquad R_i \, \psi = 0, \quad i = 1, 2, 3,$$

using the notation from Chap. 6. The solutions are given by

$$\psi(\{\alpha\}, \{\beta\}) = \mathcal{N} \left\langle T^{\alpha_1 \alpha_2 \cdots \alpha_p}_{\beta_1 \beta_2 \cdots \beta_q} \Big| \hat{g} \Big| T^{33 \cdots 3}_{33 \cdots 3} \right\rangle$$

where we use the tensor notation $T^{\alpha_1 \alpha_2 \cdots \alpha_p}_{\beta_1 \beta_2 \cdots \beta_q}$ to denote $SU(3)$ representations. (The indices $\{\beta\}$ correspond to fundamental representation and $\{\alpha\}$ to the conjugate.) Here $p - q = n$ and the lowest Landau level has $q = 0$. The energy eigenvalues are

$$E = \frac{1}{4mr^2} (R_{+i}R_{-i} + R_{-i}R_{+i}) = \frac{1}{2mr^2} \left[C_2(q+n, q) - \frac{n^2}{3} \right]$$
$$= \frac{1}{2mr^2} (q(q+n+2) + n)$$

C_2 denotes the quadratic Casimir invariant for $SU(3)$. For the lowest Landau level, $q = 0$, we get $R_a \psi = 0$, and it is easy to check that the wave functions agree with (6.70). For more information, see [13].

Problem 10.1 Write the Wilson line for an $SU(2)$ gauge field as a path integral.

Solution: Parametrize the curve C as $x^\mu(\tau)$, $x^\mu(1) = x^\mu$, $x^\mu(0) = y^\mu$, $0 \leq \tau \leq 1$. With $A^a = A^a_\mu \dot{x}^\mu$, we do the slicing of the τ-interval into segments of length $\tau_i - \tau_{i-1} = \epsilon$, so that the Wilson line is

$$W(C) = \left[\mathcal{P} \exp \left(i \int_0^1 t_a A^a d\tau \right) \right] = \left[e^{i t_a A^a (\tau_N - \tau_{N-1})} e^{i t_a A^a (\tau_{N-1} - \tau_{N-2})} \cdots \right]$$

Insert completeness relations for the coherent states and use $\langle z'|z \rangle \approx (n+1) e^{i \mathcal{A} \epsilon}$ $\langle z'|t_a|z \rangle \approx Q_a$, $z' = z + \epsilon \dot{z}$, to write

$$\langle z' | e^{i t_a A^a \epsilon} |z \rangle \approx (n+1) e^{i \int (\mathcal{A} + Q_a A^a) d\tau} \tag{12.12}$$

This leads to the path-integral expression

$$W_{kl}(C) = \int [Dz] \, \psi_k^*(z) \, e^{i \mathcal{S}(z, z')} \, \psi_l(z')$$

$$\mathcal{S}(z, z') = i \frac{n}{2} \int d\tau \, \mathrm{Tr} \left[\sigma_3 g^{-1} \left(\frac{\partial g}{\partial \tau} + A \cdot \dot{x} \, g \right) \right] \tag{12.13}$$

$$[Dz] = \frac{1}{\pi} \frac{d^2 z_N}{(1 + z_N \bar{z}_N)^2} \prod_{i=1}^{N-1} \frac{(n+1) d^2 z_i}{(4\pi \epsilon)^2 (1 + z_i \bar{z}_i)^2}$$

with $N \to \infty$, $\epsilon \to 0$ as usual.

Problem 10.2 The dynamics of a particle moving on Anti-de Sitter space in 4+1 dimensions ($= SO(4, 2)/SO(4, 1)$) can be described by the co-adjoint orbit action

$$S = \int dt \left[-i \frac{mR}{2} \mathrm{Tr}(\gamma_0 g^{-1} \dot{g}) + \frac{s_1}{2} \mathrm{Tr}(\gamma_1 \gamma_2 \, g^{-1} \dot{g}) + \frac{s_2}{2} \mathrm{Tr}(\gamma_3 \gamma_5 \, g^{-1} \dot{g}) \right]$$

where γ_μ, $\mu = 0, 1, 2, 3, 5$, are the usual Dirac matrices and s_1, s_2 label the two spins needed since $SO(4)$ is of rank 2. The group element may be parametrized as

$$g = \begin{pmatrix} \sqrt{z} & iX/\sqrt{z} \\ 0 & 1/\sqrt{z} \end{pmatrix} \Lambda, \qquad X = x^0 - \text{œ} \cdot \mathbf{x}, \qquad \Lambda = e^{-[\gamma_\mu, \gamma_\nu] \Theta^{\mu\nu}}$$

Identify the analog of (10.46) for this case and formulate fluid dynamics for mass and spin flows on AdS$_5$.

Solution: This will be left as a challenge problem.

References

1. The C^*-algebraic approach has been recently reviewed in: N.P. Landsman, *Lecture Notes on C*-algebras, Hilbert C*-modules and Quantum Mechanics*, arXiv:math-phy/9807030; C.J. Fewster and K. Rejzner, *Algebraic Quantum Field Theory- An Introduction*, arXiv:1904.04051. [hep-th].
2. Functional integration applied to quantum field theory is discussed in many books by now. A very good book, with many early references, is J. Glimm and A. Jaffe, *Quantum Physics: A Functional Integral Point of View*, Springer-Verlag (1981 & 1987). The early work which is a standard reference, particularly for fermions, is F.A. Berezin, *The Method of Second Quantization*, Academic Press (1966).
3. Path integrals originated with Dirac's discussion of the role of action in quantum mechanics. It became the basis for a quantization scheme in R.P. Feynman, Rev. Mod. Phys. 20, 367 (1948). General books include: R.P. Feynman and A.R. Hibbs, *Quantum Mechanics and Path Integrals*, McGraw Hill (1965); L.S. Schulman, *Techniques and Applications of Path Integration*, John Wiley and Sons, Inc. (1981); H. Kleinert, *Path Integrals in Quantum Mechanics, Statistics, Polymer Physics, and Financial Markets*, World Scientific Pub. Co., 3rd edition (2004).
4. V. I. Arnold, *Mathematical Methods of Classical Mechanics*, Springer-Verlag, New York (1978); V. Guillemin and S. Sternberg, *Symplectic Techniques in Physics*, Cambridge University Press (1990); J.V. José and E.J. Saletan, *Classical Dynamics: A Contemporary Approach*, Cambridge University Press (1998).
5. N.M.J. Woodhouse, *Geometric Quantization*, Clarendon Press (1992); J. Sniatycki, *Geometric Quantization and Quantum Mechanics*, Springer-Verlag (1980); S.T. Ali and M. Englis, *Quantization Methods: A Guide for Physicists and Analysts*, Rev. Math. Phys. 17, 391 (2005); A. Echeverria-Enriquez, M.C. Munoz-Lecanda, N. Roman-Roy and C. Victoria-Monge, *Mathematical Foundations of Geometric Quantization*, Extracta Mathematicae 13, 135–238 (1998).
6. V.P. Nair, *Quantum Field Theory: A Modern Perspective*, Springer-Verlag (2005).
7. An excellent new book is: P. Woit, *Quantum Theory, Groups and Representations: An Introduction*, Springer (2017).
8. For some nuances of metaplectic structure, see E. Gozzi and M. Reuter, J. Phys. A 26 6319 (1993); M. Reuter, Int. J. Mod. Phys. A 10, 65 (1995). See also our discussion in Chapter 12.
9. Coherent states are discussed in many papers and books on geometric quantization. Some general references include: A. Perelomov, *Generalized Coherent States and Their Applications*, Springer-Verlag (1986); S.T. Ali, J.-P. Antoine J. P. Gazeau and U.A. Mueller, *Coherent States*

© The Editor(s) (if applicable) and The Author(s) 2024
V. P. Nair, *Geometric Quantization and Applications to Fields and Fluids*,
SpringerBriefs in Physics, https://doi.org/10.1007/978-3-031-65801-3

and their Generalizations: A Mathematical Overview, Rev. Math. Phys. 7, 1013 (1995); J. H. Rawnsley, *Coherent States and Kähler Manifolds*, Quart. J. Math., 28, 403 (1977).

10. The Wong action is given in S. K. Wong, Nuovo Cim. A 65, 689 (1970).
11. The use of the coadjoint orbits for dynamics in the presence of monopoles, etc. is discussed in A.P. Balachandran, G. Marmo and A. Stern, Nucl. Phys. B162, 385 (1980); A.P. Balachandran, G. Marmo, A. Stern and B.S. Skagerstam, Phys. Lett. B 89, 1991 (1980); A. P. Balachandran, G. Marmo, B-S. Skagerstam and A. Stern, *Gauge Symmetries and Fibre Bundles*, Lecture Notes in Physics 188, Springer-Verlag (1982).
12. F.D. Haldane, Phys. Rev. Lett. 51 (1983) 605.
13. D. Karabali and V.P. Nair, Nucl. Phys. B 641, 533 (2002); J. Phys. A 39, 12735 (2006).
14. The index theorems we use, to the extent they are needed, can be taken from T. Eguchi, P.B. Gilkey, A.J. Hanson, Phys. Reports 66, 213 (1980).
15. For the Wess-Zumino term and anomalies, see J. Wess and B. Zumino, Phys. Lett. B 37, 95 (1971); B. Zumino, Les Houches Lectures, 1983, reprinted in S.B. Treiman et al, *Current Algebra and Anomalies*, Princeton University Press (1986); R. Stora, Lectures at the Cargèse Summer Institute on *Progress in Gauge Field Theory*, 1983. The following books may also be useful: A.P. Balachandran, *Classical Topology and Quantum States*, World Scientific Pub. Co. (1991); R.A. Bertlmann, *Anomalies in Quantum Field Theory*, Clarendon Press (1996).
16. E. Witten, Nucl. Phys. B 223, 422 (1983)
17. S.P. Novikov, Usp. Mat. Nauk. 37, 3 (1982); Russian Math. Surveys 37:5, 1 (1982).
18. The Chern-Simons term is due to S.S. Chern and J. Simons, Ann. Math. 99, 48 (1974). It was introduced into physics literature by R. Jackiw and S. Templeton, Phys. Rev. D 23, 2291 (1981); J. Schonfeld, Nucl. Phys. B 185, 157 (1981); S. Deser, R. Jackiw and S. Templeton, Phys. Rev. Lett. 48, 975 (1982); Ann. Phys. 140, 372 (1982).
19. The CS theory has applications in a wide variety of physical and mathematical problems. For relation to knot theory, see E. Witten, Commun. Math. Phys. 121, 351 (1989). Applications to the quantum Hall effect will be discussed and referenced in Chapter 9.
20. For a general formulation of the configuration space for gauge theories, and the role of reducible connections, see I.M. Singer, Commun. Math. Phys. 60, 7 (1978); M.S. Narasimhan and T.R. Ramadas, Commun. Math. Phys. 67, 121 (1979). Some other relevant papers are: I.M. Singer, Physica Scripta T24 (1981) 817; T. Killingback and E.J. Rees, Class. Quant. Grav. 4, 357 (1987); M. Atiyah, N. Hitchin and I.M. Singer, Proc. Roy. Soc. Lond. A 362, 425 (1978); P.K. Mitter and C.M. Viallet, Phys. Lett. B 85, 246 (1979); Commun. Math. Phys. 79, 457 (1981); M. Asorey and P.K. Mitter, Commun. Math. Phys. 80, 43 (1981); O. Babelon and C.M. Viallet, Commun. Math. Phys. 81, 515(1981); Phys. Lett. B 103, 45 (1981).
21. A.P. Balachandran, L. Chandar and E. Ercolessi, Int. J. Mod. Phys. A 10, 1969 (1995).
22. A. Agarwal, D. Karabali and V.P. Nair, Phys. Rev. D 96, 125008 (2017).
23. Some of the early papers on the Hamiltonian quantization are: M. Bos and V.P. Nair, Phys. Lett. B 223, 61 (1989); Int. J. Mod. Phys. A 5, 959 (1990) (we follow this work, mostly); S. Elitzur, G. Moore, A. Schwimmer and N. Seiberg, Nucl. Phys. B 326, 108 (1989); J.M.F. Labastida and A.V. Ramallo, Phys. Lett. B 227, 92 (1989); H. Murayama, Z. Phys. C 48, 79 (1990); A.P. Polychronakos, Ann. Phys. 203, 231 (1990); T.R. Ramadas, I.M. Singer and J. Weitsman, Comm. Math. Phys. 126, 409 (1989); A.P. Balachandran, M. Bourdeau and S. Jo Mod. Phys. Lett. A 4, 1923 (1989); G.V. Dunne, R. Jackiw and C.A. Trugenberger, Ann.Phys. 149, 197 (1989).
24. Geometric quantization of the Chern-Simons theory is discussed in more detail in S. Axelrod, S. Della Pietra and E. Witten, J. Diff. Geom. 33, 787 (1991).
25. E. Witten, Commun. Math. Phys. 92, 455 (1983).
26. A.M. Polyakov and P.B. Wiegmann, Phys. Lett. B 141 (1984) 223; see also D. Gonzales and A.N. Redlich, Ann. Phys.(N.Y.) 169, 104 (1986); B.M. Zupnik, Phys. Lett. B 183, 175 (1987).
27. A.P. Polychronakos, Phys. Lett. B 241, 37 (1990).
28. Most books on quantum field theory will also discuss instantons and θ-vacua from different but related points of view. See, for example, M.E. Peskin and D.V. Schroeder, *An Introduction to Quantum Field Theory*, CRC Press Taylor & Francis (1995); S. Weinberg, *The Quantum*

Theory of Fields: Volume II Modern Applications, Cambridge University Press (1996); M. Srednicki, *Quantum Field Theory*, Cambridge University Press (2007); see also Reference [6].

29. Among the original articles on θ-vacua, some of the early references are: C.G. Callan, R. Dashen and D. Gross, Phys. Lett. B 63, 334 (1976); R. Jackiw and C. Rebbi, Phys. Rev. Lett. 37, 172 (1976); R. Jackiw, Rev. Mod. Phys. 49, 681 (1977). If the spatial manifold is not simply connected one may have more vacuum angles; see, for example, A.R. Shastri, J.G. Williams and P. Zwengrowski, Int. J. Theor. Phys. 19, 1 (1980); C.J. Isham and G. Kunstatter, Phys. Lett. B 102, 417 (1981); J. Math. Phys. 23, 1668 (1982).

30. There is an enormous amount of literature on the quantum Hall effect; a lot of it is textbook material, see R.E. Prange and S.M. Girvin, *The Quantum Hall Effect*, Springer-Verlag (1987); Z.F. Ezawa, *Quantum Hall Effects: Field Theoretical Approach and Related Topics*, World Scientific (2008). Among the early papers relevant to our analysis are B.Halperin, Phys. Rev. Lett. 52, 1583 (1984); D. Arovas, R. Schrieffer and F. Wilczek, Phys. Rev. Lett. 53, 722 (1984). The description in terms of effective theory is developed in S.C. Zhang, T. Hansson and S. Kivelson, Phys. Rev. Lett. 62, 82 (1989); S. Girvin and A. MacDonald, Phys. Rev. Lett. 58, 1252 (1987); X.G. Wen and A. Zee, Nucl. Phys. B326, 619 (1989); Nucl. Phys. Proc. Suppl. 15, 135 (1990).

31. Anyons have been around in physics literature for a while; for early work, see E. Merzbacher, Am. J. Phys. 30, 237 (1960); J. Leinaas and J. Myrheim, Nuovo Cimento 37, 1 (1977); G. Goldin, R. Menikoff and D. Sharp, J. Math. Phys. 21, 650 (1980); ibid. 22, 1664 (1981); F. Wilczek, Phys. Rev. Lett. 49, 957 (1982); F .Wilczek and A. Zee, Phys. Rev. Lett. 51, 2250 (1983). For a review, see F. Wilczek, *Fractional Statistics and Anyon Superconductivity*, World Scientific, Singapore (1990). For anyons in the quantum Hall system, see R.B. Laughlin, Phys. Rev. Lett. 50, 1395 (1983); see also various articles in *Physics and Mathematics of Anyons*, S.S. Chern, C.W. Chu and C.S. Ting (eds.), World Scientific, Singapore (1991).

32. B. Binegar, J. Math. Phys. 23, 1511 (1982); R. Jackiw and V.P. Nair, Phys. Rev. D 43, 1933 (1991).

33. For a recent review, see R. Jackiw, V.P. Nair, S.Y. Pi and A.P. Polychronakos, J. Phys. A 37, R327 (2004) [arXiv:hep-ph/0407101].

34. B. Bistrovic, R. Jackiw, H. Li, V.P. Nair, and S.Y. Pi, Phys. Rev. D 67, 025013 (2003) [hep-th/0210143].

35. R. Jackiw, V.P. Nair and S.-Y. Pi, Phys. Rev. D 62, 085018 (2000).

36. V.P. Nair, Rashmi Ray and Shubho Roy, Phys. Rev. D 86, 025012 (2012).

37. D. Karabali and V.P. Nair, Phys. Rev. D 90, 105018 (2014) [arXiv:1406155].

38. D.T. Son and P. Surowka, Phys. Rev. Lett. 103, 191601 (2009).

39. G. Monteiro, A. Abanov and V.P. Nair, Phys. Rev. D 91, 125033 (2015) [arXiv:1410.4833].

40. G. 't Hooft, *Naturalness, chiral symmetry, and spontaneous chiral symmetry breaking*, NATO Adv. Study Inst. Ser. B Phys. 59, 135 (1980); *Under the spell of the gauge principle*, Adv. Ser. Math. Phys. 19, 1 (1994).

41. E. Witten, Nucl. Phys. B 223, 422 (1983); O. Kaymakcalan, S. Rajeev and J. Schechter, Phys. Rev. D 30, 594 (1984).

42. D. Kharzeev, Phys. Lett. B 633, 260 (2006); D. Kharzeev and A. Zhitnitsky, Nucl. Phys. A 797, 67 (2007); D.E. Kharzeev, L.D. McLerran and H.J. Warringa, Nucl. Phys. A 803, 227 (2008); K. Fukushima, D.E. Kharzeev and H.J. Warringa, Phys. Rev. D 78, 074033 (2008); D.E. Kharzeev, Annals Phys. 325, 205 (2010); A.V. Sadofyev and M.V. Isachenkov, Phys. Lett. B 697, 404 (2011); S. Pu, J.-H. Gao and Q. Wang, Phys. Rev. D 83, 094017 (2011); D.T. Son and P. Surowka, Phys. Rev. Lett. 103, 191601 (2009); A.V. Sadofyev, V.I. Shevchenko and V.I. Zakharov, Phys. Rev. D 83, 105025 (2011); S. Dubovsky, L. Hui, A. Nicolis and D.T. Son, arXiv:1107.0731 [hep-th]; R. Loganaygam, arXiv:1106.0277 [hep-th]; T. Kalaydzhyan and I. Kirsch, Phys. Rev. Lett. 106, 211601 (2011).

43. D. Capasso, V.P. Nair and J. Tekel, Phys. Rev. D 88, 085025 (2013).

44. F.A. Berezin, Math. USSR Izv. 6, 1117 (1972); Soviet Math. Dokl. 14, 1209 (1973); Commun. Math. Phys. 40, 153 (1975); J. Rawnsley, M. Cahen and S. Gutt, J. Geom. Phys. 7, 45 (1990); X. Ma and G. Marinescu, J. Geom. Anal. 18, 565 (2008).

45. M. Bordenmann, E. Meinrenken and M. Schlichenmaier, Commun. Math. Phys. 165, 281 (1994).; A.V. Karabegov, Trans. Amer. Math. Soc. 350, 1467 (1998); J.E. Andersen, Quantum Topol. 3, 293 (2012). A recent comprehensive review is M. Schlichenmaier, Contemp. Math. 583, 257 (2012).
46. V.P. Nair, Phys. Rev. D 102, 025015 (2020).
47. E.C.G. Sudarshan, Phys. Rev. Lett. 10, 277 (1963); C.L. Mehta, Phys. Rev. Lett. 18, 752 (1967); for reviews, see also [9].
48. D. Karabali, V.P. Nair and S. Randjbar-Daemi, *Fuzzy Spaces, the M(atrix) Model and the Quantum Hall Effect*, in *From Fields to Strings: Circumnavigating Theoretical Physics*, M. Shifman and A. Vainshtein (eds.), World Scientific Publishing Co. Pte. Ltd, Singapore, 2004; pp. 831–876.
49. G. Ishiki, T. Matsumoto and H. Muraki, Phys. Rev. **D 98**, 026002 (2018); G. Ishiki and T. Matsumoto, Prog. Theor. Exp. Phys. 2020, 013B04.
50. I. Lyris, P. Lykourgias and A.I. Karanikas, Lett. Math. Phys. 111:25 (2021).

Index

© The Editor(s) (if applicable) and The Author(s) 2024
V. P. Nair, *Geometric Quantization and Applications to Fields and Fluids*,
SpringerBriefs in Physics, https://doi.org/10.1007/978-3-031-65801-3